GENETICS

Alice Marcus

Reader
Department of Zoology
Holy Cross College
Trichirappalli
Tamil Nadu

MJP PUBLISHERS

MJP PUBLISHERS

© Publishers, 2024
All rights reserved
Printed and bound in India

47, Nallathambi Street
Triplicane
Chennai 600 005

To

*All my students who have sustained
my interest and enthusiasm in teaching.*

PREFACE

During my three decades of formal teaching, I have observed the dynamic growth of genetics and its increasing importance in the field of modern biology. But there is a body of knowledge in genetics that every biology student should be familiar with. I encourage all my students to learn the necessary details and concepts, and obtain self-learning skills rather than losing track by reading welter of details about all areas in genetics.

The impetus for writing this text comes largely from both current and former students and is designed to provide solid foundation in genetics.

My primary goal in writing this book is to provide students a reasonable full treatment of genetics in a condensed form. The book is by no means exhaustive but gives good coverage of a broad range of topics both basic and applied, and is self-sufficient. It equips the reader with necessary information with familiar examples and helpful figures. The integration of the different chapters of the book reveals how diverse concepts relate to each other.

I thank The Almighty for His immense grace and blessings without which this book would not have seen the light of day.

This book would never have been written without the whole-hearted support of my family. I wish to thank my husband Dr. Marcus Diepen Boominathan, The Principal, Bishop Heber College, Trichirappalli, my daughter Ms. Daffodil Marcus, Cognizant Solutions, Chennai, and my son Mr. Charles Marcus, CMC, Vellore, for their support and encouragement, their patience and understanding over the long durations that I spent hovering in front of the computer screen.

I record my sincere thanks to MJP Publishers, Chennai, for publishing this book.

Alice Marcus

GENETICS

CONTENTS

1

INTRODUCTION TO GENETICS

Objectives

- ✿ To look at the importance and growth of Genetics
- ✿ To gain an overview of the important discoveries in the field of Genetics
- ✿ To peruse the branches of Genetics

Key Terms

genes	pangenesis	preformationism
alleles	cell types	chromosomes
multiple factor	mitosis	meiosis
mutation	protein	phenotype
genotype	blending inheritance	germplasm theory
evolution	inheritance of acquired characters	

INTRODUCTION

Genetics is the science that studies the principles and mechanics of heredity, or the means by which traits are passed from parents to offspring. It deals with heredity and transfer of genetic information. Genes are fundamental to what and who we are. Genes are bits of biochemical instructions found inside the cells of every organism from bacteria to humans. Offspring receive a mixture of genetic information from both parents. This process contributes to the great variation of traits that we often see in nature, such as the colour pattern in butterfly wings or markings on a flower petal.

Geneticists work to understand how the encoded information in the genes is expressed and transmitted from generation to generation. Research findings in Genetics are valuable and have ethical and economic implications and are of importance to individual, family and society at large. Scientists uncover the mystery behind how our genes work and what they can reveal to us—the possibility of having certain diseases and conditions. The scientific field of genetics can help families affected by genetic disorders to have a better understanding about heredity, what causes various genetic disorders to occur, and what possible strategies can be used to decrease the incidence of genetic disorders.

The discovery of genes that influence our traits and lives; susceptibility to diseases; cancer-causing genes; gene therapy and treatment of diseases are regularly reported that make the study of genetics interesting to any one and more to a biology student. An understanding of Genetics is critical to the students of biology. It provides the unifying principles in biology: all organisms encode genetic information in the same way and use nucleic acids. Genetics is the basis for various biological disciplines. Developmental biology heavily relies on regulated gene expression. Evolution requires the study of genetic change taking place through time. Genetic variation is vital in understanding the past, present and future of life. The study of taxonomy, ecology and animal behaviour is making use of genetic methods. The study of biology and medicine is incomplete without the understanding of genes and their role.

The application of Genetics was evidenced in the Green Revolution in the 1960s that saw revolution in food production. Biotechnology industry employs molecular genetic techniques to develop and produce in large scale the commercially valuable products. Genetically engineered crops and pharmaceuticals are a reality today. Genetics also play an important role in medicine. It is recognized that almost all diseases and disorders have hereditary component. Molecular genetics has allowed important insights into the nature of cancer and development of diagnosis and treatment. The direct alteration of genes to treat hereditary disease—gene therapy—is a revolution in the treatment of genetic diseases.

HISTORY

Although Genetics is relatively a new branch of science, the hereditary nature of traits have been understood and practised for thousands of years and the genetic principles have been applied in the domestication of plants and animals by early humans. General aspects of the problem of heredity, was known to Babylonians and Aryans thousands of years before the Christian era. In 6000 BC, Babylonians depicted the pedigree of horses and cross pollination of palm. About 4000 BC, the Chinese tried to improve the varieties of rice. The first domesticated organisms included wheat, barley, peas, dogs, goats and sheep.

Ancient writings demonstrate that early humans were aware of their own heredity. The ancient Greeks proposed the concept of Pangenesis, which proposed that particles called gemmules carry information from various parts of the body to the reproductive organs, from where they are passed to the embryo at the time of conception. This concept, though incorrect, persisted till the 1800s. Later the concept of inheritance of acquired characters was proposed by Greeks in which the traits that are acquired during one's lifetime are inherited and passed on from parents to offspring. This concept also was not accepted but remained through the twentieth century. The Greek philosopher Aristotle (384–322 BC) rejected the concepts of pangenesis and inheritance of acquired characters, pointing out that people sometimes resemble their past ancestors more than their parents and that acquired mutilated body parts are not inherited. He believed that both male and female contribute equally to the offspring.

The discovery of microscopes in 1500 and the discovery of cells by Robert Hooke in 1665 made naturalists to propose the concept of preformationism. This explains that inside the egg or sperm existed a tiny miniature adult, a *homunculus*, which simply enlarged during development. Another early idea was blending inheritance, which proposed that offspring are a mixture of parental characteristics.

Developments in cytology in the 1800s strongly influenced the study of genetics. Robert Brown discovered nucleus in 1833 and cell theory was proposed by Schleiden and Schwann in 1839. The cell theory explains that all life is composed of cells, cells arise from pre-existing cells only and the cell is the fundamental unit of structure and function in living organisms. Biologists started to examine cells to find out the transmission of traits through cell division. Walter Flemming observed the division of chromosomes in 1879 and described mitosis. In 1885 it was recognized that nucleus contained the hereditary information. August Weismann proposed the germplasm theory and explained that the reproductive organs carry the complete set of genetic information that is passed to the gametes, the reproductive cells.

The first marked pioneer and experimental work in Genetics was started by Gregor Johann Mendel in 1866. His discoveries provided the foundation for the science of Genetics. Mendel's paper was published in the Proceedings of the Brunn Society for Natural History. It explained the results of his hybridization experiments conducted with garden pea *Pisum sativum*. This valuable contribution was unnoticed. But it was established as a distinct branch when Mendel's findings were rediscovered by three botanists—Hugo De Vries from Holland, Carl Correns from Germany and Erich Von Tshermak from Austria independently in 1900.

Today, Genetics is a mature but dynamic science, recognized as the very core of modern biology. The Science of Genetics was founded by Mendel and obtained the present status by the contributions of a large number of scientists, which are tabulated in the following section.

Year	Discovery
1901	Hugo De Vries coined the term "mutation" to describe sudden, spontaneous changes in *Oenothera* plant.
1902	Boveri and Sutton demonstrated the occurrence of paired chromosomes (homologous) in diploid species. C.E. McClung described sex chromosomes in grasshopper. W.S. Sutton proposed the chromosome theory of heredity.
1906	Bateson named the science, Genetics. Bateson and R.C. Punnett first reported linkage in sweet pea.
1908	Hardy and Weinberg formulated the "Hardy–Weinberg Law" relating gene frequencies and genotype frequencies in randomly mating populations. H. Nilsson Ehle put forwarded the multiple factor hypothesis.
1909	W. Johannsen coined the term "gene". Garrod published the book *"Inborn Errors of Metabolism"*.
1910	Thomas Hunt Morgan (Nobel Prize, 1933) established sex-linkage in *Drosophila*. *Drosophila* genetics began.
1917–23	C.B. Bridges described different types of chromosomal abnormalities and aberrations. He explained genic balance theory and non-disjunction.
1925	Bernestein suggested the multiple allelic inheritance of ABO blood group.
1927	H.J. Muller (Nobel Prize, 1946) reported the use of ClB technique to demonstrate that X-rays are mutagenic.

Year	Discovery
1928	Griffith discovered transformation in *Diplococcus pneumoniae*.
1930	Ronald A. Fisher, John B.S. Haldane and Sewall Wright laid the foundation for population genetics.
1931	S. Stern provided cytological proof of crossing over in *Drosophila*.
1940	K.S. Landsteiner and A.S. Wiener discovered Rh factor.
1941	Beadle and Tatum (Nobel Prize, 1958) established "one gene–one enzyme" concept in *Neurospora*.
1944	O.T. Avery, C.M. MacLeod and M. McCarty demonstrated the transforming principle as DNA. Lederberg (Nobel Prize, 1958) and Tatum discovered conjugation in bacteria.
1950	E. Chargaff demonstrated that in DNA the number of adenine groups is equivalent to thymine groups and the number of cytosine groups is equal to guanine groups.
1952	Hershey (Nobel Prize, 1969) and Chase showed that the genetic material of bacteriophage is DNA. Zinder and Lederberg discovered phage-mediated transduction in bacteria.
1953	J.D. Watson and F.H.C. Crick (Nobel Prize, 1962) proposed the double-helix structure of DNA using X-ray diffraction data of Wikins (Nobel Prize, 1962) and the base composition data of Chargaff.

Year	Discovery
1955	Benzer described the fine structure of the phage T4rII locus.
1956	Tjio and Levan established the normal diploid chromosome number in humans to be 46.
1957	Fraenkel Conrat and Singer demonstrated that the genetic material of tobacco mosaic virus is RNA.
1958	Meselson and Stahl explained semi-conservative replication of DNA.
	Ingram showed that normal and sickle-cell haemoglobin differ in a single amino acid residue.
	Kornberg (Nobel Prize, 1959) isolated DNA polymerase I from *E.coli*.
1959	Ochoa (Nobel Prize, 1959) discovered the RNA polymerase first.
1961	M.W. Nirenberg and J.H. Mathaei cracked the genetic code present on mRNA.
	Jacob and Monod (Nobel Prize, 1965) proposed the "Operon Model" for gene regulation.
1965	Harries Watkins formed hybrid cell by the fusion of somatic cells of mouse and man.
	Holley (Nobel Prize, 1968) worked out the first complete nucleotide sequence of a yeast alanine tRNA.
1970	Nathan and Smith (Nobel Prize, 1978) isolated the first restriction endonucleases.
	Baltimore (Nobel Prize, 1975) identified reverse transcriptase by RNA tumour viruses.
1972	Berg (Nobel Prize, 1980) produced first recombinant DNA *in vitro*.

Year	Discovery
1976	Bishop and Varmus (Nobel Prize, 1989) demonstrated the proto-oncogene and oncogene relationship.
1977	Maxim, Gilbert and Singer (Nobel Prize, 1980) published DNA sequencing techniques.
1983	Kary Mullis and others developed the polymerase chain reaction for quickly amplifying DNA. McClintock received Nobel Prize for introducing the jumping genes concept.
1986	Nirenberg and Khorana established the complete genetic code.
1989	Tsui and Collins clone the cystic fibrosis gene.
1990	Gene therapy was used for the first time to treat human genetic disease in the United States. World's first Human Genome Project was launched.
1995	The first complete DNA sequence of a free-living organism—the bacterium *Haemophilus influenzae*—was determined.
1996	*Saccharomyces cerevisiae* is the first eukaryotic genome to be released.
1998	A rough draft of the human genome map was produced, showing the locations of more than 30,000 genes.
2001	Craig Venter and Francis Collins announced the first complete draft of the human genome to the world.
2003	Successful completion of Human Genome Project with 98% of the genome sequenced to a 99.99% accuracy.

FUTURE OF GENETICS

The history of genetics given exposes the accelerating pace of advances in genetics. The flood of new genetic information is supported by sophisticated computer programs to store, retrieve, compare and analyse genetic data, and this resulted in merging of molecular biology and computer science that has given rise to the field bioinformatics. The genome sequences of many organisms are added to DNA databases frequently.

In future, the focus of DNA sequencing will shift from the genomes of different species to individual differences within species. New genetic microchips that simultaneously analyse thousands of RNA molecules will provide information about the activity of thousands of genes in a given cell in response to external signals, environmental stresses and disease states.

The use of genetics in the agricultural, chemical and health care fields will continue to expand. This ever-widening scope of genetics will raise significant ethical, social and economic issues.

BASIC CONCEPTS IN GENETICS

✧ **Cell types** Structurally cells are of two types. Prokaryotic cells are those that lack a cell membrane and are without membrane-bound cell organelles. Eukaryotic cells are more complex, possessing a nucleus and membrane-bound organelles.

✧ **Gene** Gene is the fundamental unit of heredity. It is the unit of genetic information that encodes a characteristic. Genetic information is encoded in the nucleic acids, which are of two types namely deoxyribonucleic acid (DNA) and ribonucleic acid (RNA). They are polymers consisting of repeating units called nucleotides; each nucleotide consists of a sugar, a phosphate, and a nitrogenous base. The nitrogenous bases are purine and pyrimidine. In DNA the bases are adenine, guanine, thymine and cytosine (abbreviated A, G, T, and C). The sequence of these bases encodes genetic information. Most organisms carry genetic information in DNA, but few viruses carry it in RNA. In RNA the base thymine is replaced by uracil (U).

✪ **Alleles** A gene that specifies a characteristic may exist in many forms, called alleles.

✪ **Genotype and phenotype** The genetic information that an organism possesses is its genotype and the trait is the phenotype.

✪ **Multiple factors** Some traits are influenced by multiple genes that interact with environmental factors.

✪ **Chromosomes** Genes are located in chromosomes. The vehicles of genetic information within the cell are chromosomes, which consist of DNA and proteins. Each organism has a specific number of chromosomes; for example human has 46; mouse possesses 40 and pigeon has 80. Each chromosome carries a large number of genes.

✪ **Mitosis and meiosis** Cells multiply by cell division. Chromosomes separate through the process of mitosis and meiosis. The cell division ensures that each daughter cell receives a complete set of chromosomes. Mitosis is the separation of replicated chromosomes during the division of a somatic (non-sex) cells. Meiosis is the pairing and separation of replicated chromosomes during the division of sex cells to produce gametes.

✪ **Protein** Genetic information is transferred from DNA to RNA to proteins. Genes encode traits by specifying the structure of proteins. Genetic information is first transcribed from DNA into RNA, and then RNA is translated into the amino acid sequence of a protein.

✿ **Mutation**   Mutations are permanent, heritable changes in genetic information. They are of two major types. Gene mutations affect only the genetic information of a single gene; chromosomal mutations alter the number or structure of chromosomes and therefore generally affect many genes.

✿ **Evolution**   It is a genetic change and involves two steps. First, the genetic variation arises randomly and then these genetic variants either increase or decrease in frequency. Genetic change through time is evolution.

BRANCHES OF GENETICS

Geneticists have divided the science of genetics into the following branches:

✿ **Behavioural genetics** is the field of biology that studies the role of genetics in animal (including human) behaviour. Classically, behavioural geneticists have studied the inheritance of behavioural traits.

✿ **Biometric genetics** is concerned with the inheritance of quantitative traits (e.g. weight, height, milk production, etc.) and predicts the relationships of individuals based on given data.

✿ **Biochemical genetics** deals with biochemical aspects of heredity and is concerned with inborn errors.

✿ **Clinical genetics** deals with application of the principles of genetics in analysing and diagnosing various hereditary diseases in man and suggests possible cures for them.

✿ **Cytogenetics** is concerned with physical and physiochemical aspects of heredity.

✿ **Developmental genetics** studies the functioning of genes and their interactions in the development of organisms.

- **Ecological genetics** is the study of the interactions among organisms and between the organisms and their environment. Ecological genetics studies phenotypic evolution in natural populations of organisms.

- **Eugenics** is concerned with the application of the principles of genetics to the improvement of human race.

- **Evolutionary genetics** is the broad field of study that attempts to account for evolution in terms of changes in gene and genotype frequencies within populations. A focus of evolutionary genetics is to describe how the evolutionary forces shape the patterns of biodiversity observed in nature.

- **Genetic engineering**, recombinant DNA technology, genetic modification/manipulation (GM) are terms that apply to the direct manipulation of an organism's genes. Genetic engineering uses the techniques of molecular cloning and transformation to alter the structure and characteristics of genes directly.

- **Genomics** is the study of an organism's entire genome. The field includes intensive efforts to determine the entire DNA sequence of organisms and fine-scale genetic mapping efforts.

- **Human genetics** describes the study of inheritance as it occurs in human beings. Human genetics encompasses a variety of overlapping fields including classical genetics, cytogenetics, clinical genetics, molecular genetics, biochemical genetics, genomics, developmental genetics and genetic counselling.

- **Medical genetics** is the speciality of medicine that involves the diagnosis and management of hereditary disorders. Medical genetics refers to the application of genetics to medical care. Genetic medicine is a newer term for medical genetics and incorporates areas such as gene therapy and personalized medicine.

✿ **Microbial genetics** is the study of genetics of microorganisms (virus, bacteria, unicellular plants and animals).

✿ **Molecular genetics** is the study of gene—its structure, organization and function.

✿ **Radiation genetics** is the study of effect of radiations on genes and the change in gene expression.

Summary

✿ Genetics is the study of the gene and its transmission.

✿ It plays an important role in understanding life and variations.

✿ The history of genetics is traced from the early seventeenth century.

✿ Gregor Mendel is the pioneer in genetics.

✿ The basic concepts in genetics reveal the relations between genes, DNA and chromosomes.

✿ The dynamism of genetics is evidenced in the various branches of genetics listed.

REVIEW QUESTIONS

1. State the principles of genetics.

2. Outline the importance of genetics.

3. Trace the early history of genetics up to 1900.

4. Highlight the basic concepts in genetics.

5. List the branches of genetics.

2

NUCLEIC ACID AS THE BASIS OF INHERITANCE

Objectives

- ✪ To understand the properties of the genetic material
- ✪ To investigate the evidences to prove that DNA is the genetic material
- ✪ To examine the structure of DNA

Key Terms

DNA	RNA	replication
nucleotides	purines	complementarity
pyrimidines	phosphodiester	tobacco mosaic virus
A-DNA	B-DNA	Z-DNA
phage labelling	transforming principle	

INTRODUCTION

It has been known for a century that genes, the discrete functional units of genetic material, are located in chromosomes within the nuclei of eukaryotic cells. On account of their structural complexity, proteins were considered to be the hereditary material for a very long time, until experiments were conducted to prove that nucleic acids work as hereditary material. DNA (deoxyribonucleic acid) is found to be the genetic material in almost all the living beings except some plant viruses where RNA acts as the genetic material. In 1944, the Oswald Avery and his colleagues gave the first proof that DNA is the genetic material.

BASIC CHARACTERISTICS NECESSARY FOR GENETIC MATERIAL

The genetic material must carry information to control the synthesis of enzymes and proteins within a cell or an organism. It should self-replicate with high fidelity and be located in the chromosomes.

Protein Synthesis

DNA possesses the complexity required to direct protein synthesis. The base sequences of DNA specify a particular amino acid during protein synthesis. The genetic code gives the relationship of DNA bases to the amino acids in proteins and enzymes too.

Replication

Watson and Crick model of DNA threw light on complementarity. Since the base sequence on one strand is complementary to the base sequence on the opposite strand, each strand could act as a template for a new double helix if the molecule simply "unzipped".

Location

DNA must reside in the nucleus of eukaryotes, where the genes occur on chromosomes. In all the cells, DNA is in the chromosomes. The chromosomes are passed from cell to cell during cell division.

EVIDENCE FOR DNA AS THE GENETIC MATERIAL

Transformation

F. Griffith (1928) reported that heat-killed bacteria of one type could "transform" living bacteria of a different type. He demonstrated the process of transformation using two strains of the bacterium *Streptococcus pneumoniae*. One strain (S) produced smooth colonies on media in a Petri plate because the cells had polysaccharide capsules. It causes fatal bacteraemia (viable bacteria present in the blood) in mice. Another strain (R), which lacked polysaccharide capsules, produced rough colonies on Petri plates; it did not have a pathological effect on mice (Figure 2.1).

Griffith found that neither heat-killed S-type nor live R-type cells, by themselves caused bacteraemia in mice. But when he injected a mixture of live R-type and heat-killed S-type cells into mice, the mice developed bacteraemia. Thus, the heat-killed S-type cells had something that transformed R-type into S-type cells.

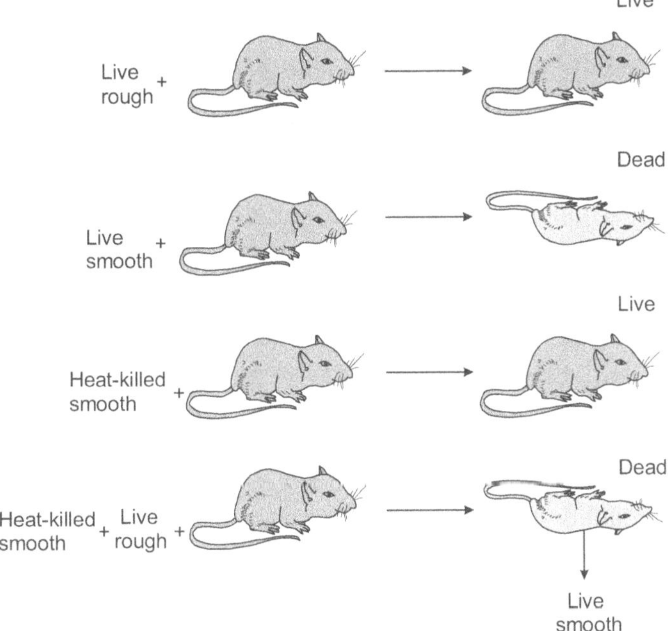

Figure 2.1   Griffith's experiment
(*Source*: www.nature.com)

Oswald Avery, C. MacLeod and M. McCarty in 1944 reported the "nature of the transforming substance". They did their work *in vitro*. They ruled out proteins, carbohydrates and lipids by their extraction procedure and the protease enzymes as well as ribonuclease enzymes had no effect on the transforming material. They demonstrated that the only enzymes that destroyed the transforming ability were enzymes that destroyed DNA. They showed that the transforming principle precipitated at the same rate as the purified DNA and it absorbed ultraviolet light at the same wavelength as DNA. This study provided the first experimental evidence that DNA is the genetic material and the "transforming principle".

Phage Labelling

In 1952, A.D. Hershey and M. Chase supported (by their experiments) the notion that DNA is the hereditary material. They designed an experiment using radioactive isotopes of sulphur and phosphorus to keep separate track of the viral proteins and nucleic acids during the injection process. They used the T2 bacteriophage and the bacterium *Escherichia coli*. The phages were labelled by radioactive isotopes ^{35}S or ^{32}P in culture medium. All nucleic acids contain phosphorus whereas proteins contain sulphur (Figure 2.2).

When ^{32}P-labelled phages were mixed with unlabelled *E. coli* cells, Hershey and Chase found that the ^{32}P label entered the bacterial cells and that the next generation of phages that burst from infected cells carried a significant amount of ^{32}P label. When ^{35}S-labelled phages were mixed with unlabelled *E. coli,* the researchers found that the ^{35}S label stayed outside the bacteria for the most part. Thus they demonstrated that the outer protein coat of a phage does not enter the bacterium it injects, whereas the phage's inner material, consisting of DNA, does enter the bacterial cell. Since the DNA is responsible for the production of the new phages during the injection process, the DNA, not the protein, must be the genetic material.

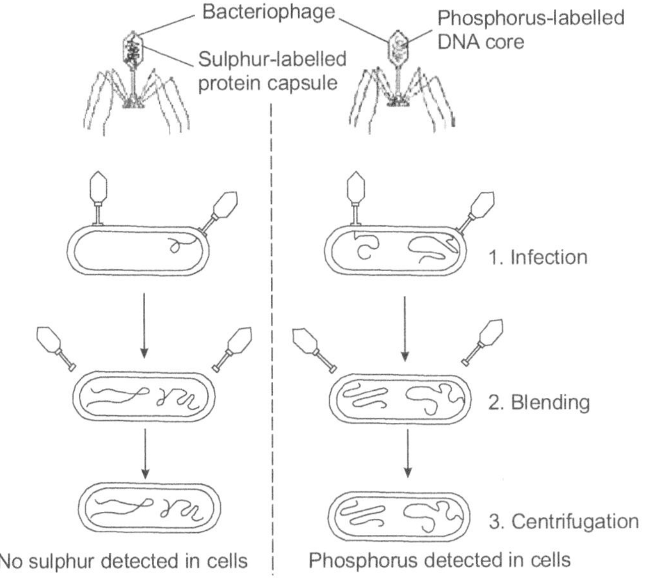

Figure 2.2   Hershey and Chase experiment
(*Source*: www.accessexcellence.org)

Base Equivalence in DNA

E. Chargaff (1943–53) made a quantitative analysis of four bases in hydrolysates of DNA specimens isolated from different organisms and put forward the following facts.

i.   Base composition of DNA varies from one species to another.

ii. DNA isolated from different tissues of the same species has the same base composition.

iii. The base composition of DNA in a given species does not change with age, nutritional state or change in environment.

iv. The number of adenine residues is always equal to the number of thymine residues (A = T) and the number of guanine residues is always equal to the number of cytosine residues (G = C). Always purine residues are equal to the sum of pyrimidine residues (A + G = T + C).

v. The DNAs extracted from closely related species have similar base composition whereas those from widely different species are likely to have widely different base composition.

RNA AS GENETIC MATERIAL

DNA is the genetic material in most organisms but in some viruses, RNA is the genetic material. Viruses like tobacco mosaic virus (TMV) consist only of RNA and protein. The TMV possesses a single molecule of RNA surrounded by a helically arranged cylinder of protein molecules. In 1955, H. Fraenkel-Conrat and B. Singer combined the RNA from one strain of TMV and protein from another strain and constituted recombinational type of TMV. The TMV thus produced is always the type which contributed the RNA.

Thus it is concluded that in viruses that do not have DNA, RNA serves as the genetic material.

CHEMICAL NATURE OF NUCLEIC ACIDS

Nucleic acids are made by joining nucleotides in a long chain. Nucleotides are made up of three components: phosphate, sugar and nitrogenous base. DNA and RNA both have four bases, two purines and two pyrimidines in their nucleotide chains. Both DNA and RNA have the purines adenine and guanine and the pyrimidine cytosine, RNA has the pyrimidine uracil and DNA has thymine (Table 2.1). The adenine is paired with thymine by a double hydrogen bond while guanine is paired with cytosine by a triple hydrogen bond. The sugars differ only in the presence (ribose in RNA) or absence (deoxyribose in DNA) of oxygen in the 2′ position (Figure 2.3).

Table 2.1   Components of nucleic acids

	Phosphate	**Sugar**	**Purines**	**Pyrimidines**
DNA	Present	Deoxyribose	Guanine Adenine	Cytosine Thymine
RNA	Present	Ribose	Guanine Adenine	Cytosine Uracil

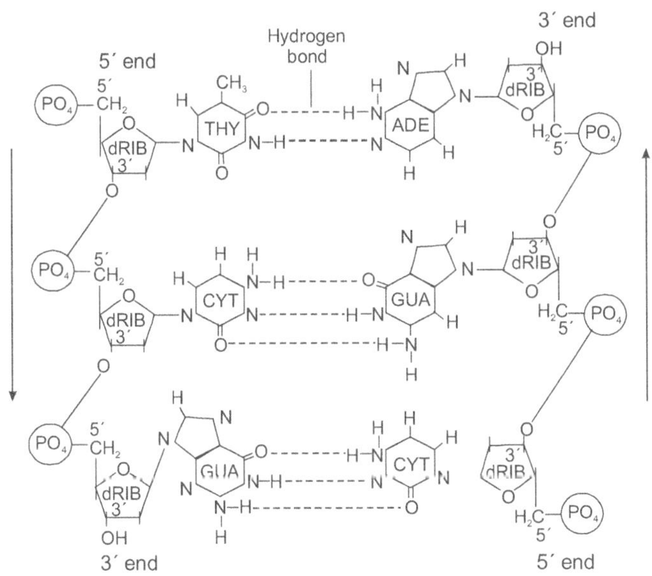

Figure 2.3   Chemical structure of DNA
(*Source*: www.academic.brooklyn.cuny.edu)

A nucleotide is formed in the cell when a base attaches to the 1′ carbon of the sugar and a phosphate attaches to the 5′ carbon of the same sugar. Nucleotides are linked together by the formation of a bond between the phosphate at the 5′ carbon of one nucleotide and the hydroxyl (OH) group at the 3′ carbon of an adjacent molecule. This is phosphodiester bonding.

ALTERNATIVE FORMS OF DNA

DNA occurs in different forms. The right-handed helix is B-DNA. It turns in a clockwise manner when viewed down its axis. The bases are stacked almost exactly perpendicular to the main axis with ten bases per turn. If the water content increases to about 75%, the A form of DNA (A-DNA) will occur. In A-DNA the bases are tilted and there are more base pairs per turn. In 1979, Alexander Rich and his colleagues discovered a left-handed helix called Z-DNA in which the backbone formed a zigzag structure.

Summary

- The characteristics necessary for the genetic material are: it must be able to replicate and synthesize protein and must be located in the chromosome.
- The evidences for DNA as the genetic material were provided by Griffth, Avery *et al.*, Hershey and Chase, and Chargaff.
- In some viruses the genetic material is RNA. The double-helix structure of DNA was proposed by Watson and Crick.
- DNA is made up of nucleotides composed of purines, pyrimidines, deoxyribose, and phosphates.
- DNA can exist in alternative forms as B-DNA, A-DNA and Z-DNA.

REVIEW QUESTIONS

1. Enumerate the basic characteristics necessary for the genetic material.

2. Explain the contributions of Griffith and Avery *et al.* to the study of genetic material.

3. Cite the evidences provided by Hershey and Chargaff in favour of DNA as the genetic material.

4. Draw the chemical structure of DNA showing three base pairs.

5. Distinguish the B-DNA from A-DNA and Z-DNA.

3

MENDELIAN GENETICS—BASIC PRINCIPLES OF HEREDITY

Objectives

- ✿ To learn the experiments conducted by Mendel
- ✿ To understand Mendel's laws of inheritance

Key Terms

monohybrid cross	pure line	dominant trait
recessive trait	alleles	homozygous
heterozygous	Punnett square	test cross
law of segregation	dihybrid cross	trihybrid cross
law of independent assortment		

INTRODUCTION

Gregor Johann Mendel, an Austrian monk, was the first one to explain the mechanism of inheritance. He is therefore called "Father of Genetics". Mendel conducted his historic experiments in garden pea plant *Pisum sativum* for about nine years. He deduced the three basic laws of inheritance of characters. These are now called Mendel's laws of inheritance. The results of his classical experiments and his explanations were published in the Journal *The Annual Proceedings of the Natural History Society of Brunn* in 1865. His outstanding contributions were overlooked by the scientific community. He died at the age of 61 in 1884, unrecognized for his contribution to genetics. But later in 1900, three botanists from different places conducted similar experiments independently and arrived at conclusions similar to those of Mendel. They were Hugo de Vries from Holland, Carl Correns from Germany and Eric Von Tschermak from Austria. They interpreted their results in terms of Mendel's principles and drew attention to his pioneering work.

MENDEL'S EXPERIMENTS

Mendel's approach to the study of genetics was successful because of his choice of the pea plant for his experiments. He adopted an experimental approach. He formulated hypotheses on his initial observations and then conducted repeated crosses to test his hypotheses. He kept a careful record of all his observations and results in detail. He spent almost ten years to conduct experiments before publishing.

Mendel selected the garden pea plant *Pisum sativum* for the following reasons.

- ✿ It has a short life cycle and so it is possible to study several generations within a short period.
- ✿ Self-pollination is the predominant mode of reproduction.

Table 3.1   List of traits selected by Mendel

Character	Dominant	Recessive
Stem length	Tall	Dwarf
Flower position	Axial	Terminal
Pod shape	Inflated	Constricted
Pod colour	Green	Yellow
Seed shape	Round	Wrinkled
Seed colour	Yellow	Green
Seed coat colour	Grey	White

✪   Because of self-fertilization, it is easy to get pure lines for several generations.

✪   It is easy to cross by transferring pollen from one plant to another.

Mendel chose seven pairs of non-overlapping characters of the pea plant for study as listed in Table 3.1.

MONOHYBRID CROSS

Mendel began by studying monohybrid crosses—those between parents that differed in a single characteristic. A single pair of contrasting characters was selected for conducting experiments.

Mendel selected pure line of tall and dwarf plants (that is, the tall plants always produced tall plants and dwarf plants always produced dwarf plants). These plants represented parental (P_1) generation. The plants raised by sowing the seeds from these parents (P_1) belonged to first filial (F_1) generation. The plants raised from the seeds of F_1 belonged to second filial (F_2) generation.

When a pure tall (TT) plant was crossed with a pure dwarf (tt) plant, in the F_1 generation all the plants were found to be tall. When the F_1 plants were crossed among themselves, in the F_2 generation both tall and dwarf plants were produced in the ratio of 3 : 1. The dwarf plant of F_2 produced only dwarf plant while among tall plants, 1/3 produced only tall plant and 2/3 produced tall and dwarf in the ratio of 3 : 1 in the F_3 generation. It means F_2 generation consisted of three types of plants namely,

i.   Tall (pure)   25%

ii.   Tall (hybrid) 50%

iii.   Dwarf (pure) 25%

The results of the above experiment can be explained as follows:

- ✿ Tallness and dwarfness are determined by a pair of contrasting factors or determiners (now called genes). A plant is tall because it has gene (T) for tallness and since a trait is represented always by a pair of genes present in the homologous chromosomes that are received one from either parent, the tallness is represented as TT.

- ✿ When a tall and dwarf plant are crossed, in the F_1 generation only tallness is expressed. Tallness masked the expression of dwarfness. So tallness is called a dominant trait and dwarfness, a recessive trait.

A pair of genes, representing the two alternatives of the same character and located in the same locus in the homologous chromosomes are called alleles or allelomorphs. In the gene pair Tt, controlling the height of the plant, 'T' is present on one chromosome and 't' on the other homologous chromosome.

- ✿ The genes are never contaminated. When gametes are formed, these factors separate so that each gamete gets only one of the two alternative factors. Tallness (T) and dwarfness (t) are separate entities and in a gamete only one of the two that is either T or t is present. They separate out and unite independently to produce tall and dwarf plants.

Every organism possesses two genes for every character. If in an organism the two genes for a particular character are identical, it is said to be pure or homozygous for that character. The tall plant is TT and dwarf plant is tt. On the other hand heterozygous (hybrid) organism possesses contrasting genes of a pair. It receives two different alleles for the same trait from its two parents (Figure 3.1). Tt is heterozygous tall plant producing two different types of gametes—one carrying T gene and the other t.

☼   The term genotype designates the genetic make-up of an organism, whereas phenotype is the appearance of an individual. For example, in F_2 generation the tall plants are TT (homozygous) and Tt (hybrid). But irrespective of their genetic composition they develop into tall plants. So externally, that is, phenotypically, they are tall and genotypically they are TT or tt.

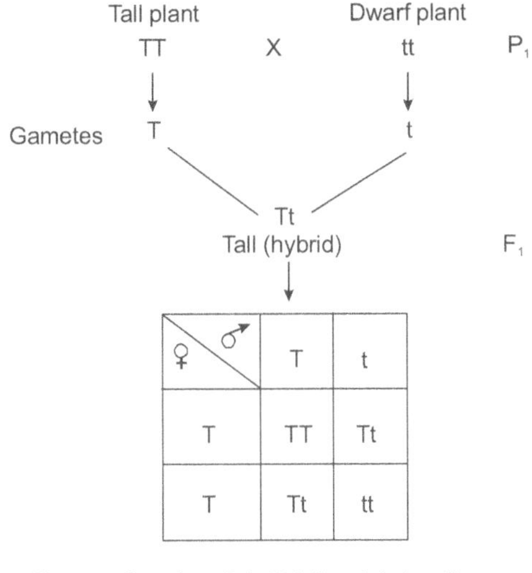

Phenotypic ratio = 3:1 (3 tall and 1 dwarf)

Genotypic ratio = 1:2:1 (1 homozygous tall
2 heterozygous tall
1 dwarf)

Figure 3.1   Monohybrid cross

The Punnett Square

The outcome of the genetic crosses has been predicted by a simple shorthand method called Punnett square. This is constructed by drawing a grid, putting the gametes produced by one parent along the upper edge and the gametes produced by the other parent down the left side. Each cell (a block within) contains an allele from each of the corresponding gametes, generating the genotype of the progeny produced by fusion of those gametes.

Test Cross and Backcross

Crosses between F_1 offspring with either of the two parents are known as backcrosses. When F_1 offspring are crossed with the dominant parents, all the F_2 offspring develop dominant characters. On the other hand when F_1 hybrids are crossed with recessive parent, individuals with both phenotypes tall and dwarf occur in equal proportion. Both the crosses are known as backcross, and the second one is specified as test cross.

Test cross is a cross between heterozygous F_1 hybrid and the recessive homozygous parent. The test cross ratio is 1:1. The test cross is used to determine whether the dominant parents are homozygous or heterozygous. A cross between F_1 hybrids of the monohybrid cross with dwarf homozygous recessive parents, produces tall and dwarf in equal proportion indicating that F_1 hybrids are heterozygous (Figure 3.2).

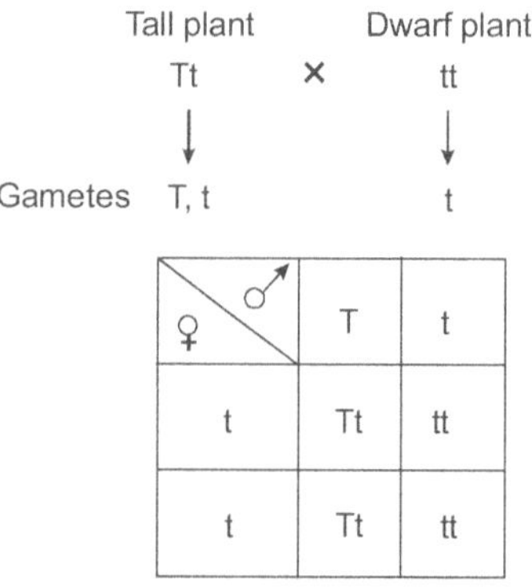

Figure 3.2   Test cross

Test cross ratio is 1:1 (50% heterozygous tall, and 50% dwarf).

Mendel's Law of Segregation

The conclusions that Mendel developed about inheritance from his monohybrid crosses have been described as the principle of segregation and the concept of dominance.

Mendel's first law namely the law of segregation states that each individual possesses two alleles for any particular characteristic. These two alleles segregate (separate) when gametes are formed, and one allele goes into each gamete. Furthermore, the two alleles segregate in equal proportions.

The concept of dominance states that, when two different alleles are present in a genotype, only the trait of one of them is observed in the phenotype and this allele is called the dominant allele.

DIHYBRID CROSS

In a dihybrid cross two pairs of characters are selected for the experiment. The cross was made between plants having round and yellow (RRYY) cotyledons and plants having wrinkled and green (rryy) cotyledons. One pair of character considered is the colour of the seed (yellow or green) and the other is texture of the seed (round or wrinkled). It is found that yellow (y) is dominant over green (Y) and round (R) is dominant over wrinkled (r). The F_1 hybrids all had round and yellow seeds and when these plants were self-fertilized, they produced four types of plants: 315 round, yellow seeds; 101 wrinkled, yellow seeds; 108 round, green seeds; and 32 wrinkled, green seeds (Figure 3.3). They appeared approximately in the following proportion in F_2 generation.

i.	Round and yellow	9/16
ii.	Wrinkled and yellow	3/16

iii. Round and green   3/16

iv. Wrinkled and green   1/16

The phenotypic dihybrid ratio is $9:3:3:1$

There are 16 different genotypes in dihybrid cross representing only 4 phenotypes. But there are 9 genotypes and the genotypic ratio is $1:2:4:2:1:2:1:2:1$

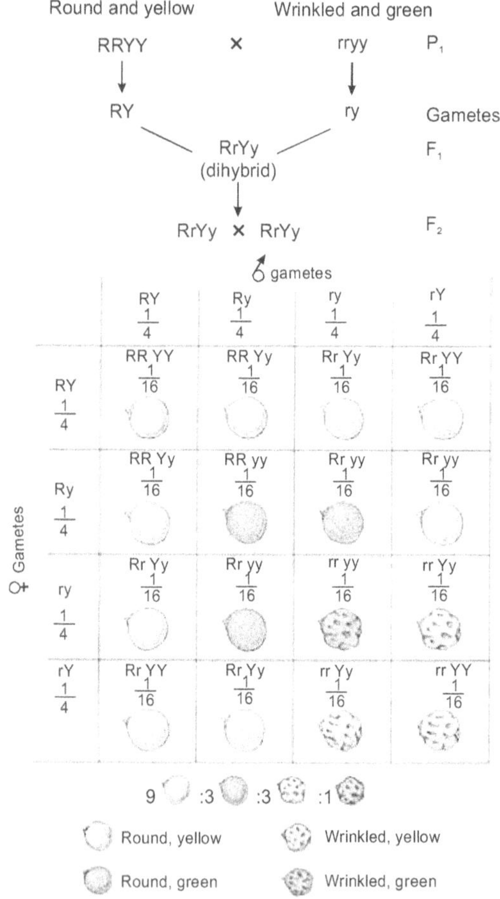

Figure 3.3   Dihybrid cross
(*Source:* www.emc.maricopa.edu)

Phenotype		**Genotype**	
Round yellow	9	RRYY	1
		RRYy	2
		RrYy	4
		RrYY	2
Round green	3	RRyy	1
		Rryy	2
Wrinkled yellow	3	rrYY	1
		rrYy	2
Wrinkled green	1	rryy	1

Based on the dihybrid cross, Mendel postulated the second law called the law of independent assortment.

Law of Independent Assortment

If the inheritance of more than one pair of characters is taken, the alleles of a character located at different loci separate and assort independently of one another.

The results of the dihybrid cross can be explained as follows. Each plant possesses two alleles coding for each characteristic, so the parental plants must have had genotypes RRYY and rryy. The law of segregation indicates that the alleles for each locus separate, and one allele for each locus passes to each gamete. The gametes produced by the round, yellow parent therefore contain alleles RY, whereas the gametes produced by the wrinkled, green parent contain ry. These two types of gametes unite to produce the F_1, all with genotype RrYy. Because round is dominant over wrinkled and yellow is dominant over green, the phenotype of the F_1 will be round and yellow.

When Mendel self-fertilized the F_1 plants to produce the F_2, the alleles for each locus separated, with one allele going into each gamete. This is where the principle of independent assortment becomes important. Each pair of alleles can separate in two ways:

1. R separates with Y, and r separates with y to produce gametes RY and ry or

2. R separates with y, and r separates with Y to produce gametes Ry and rY

The principle of independent assortment explains that the alleles at each locus separate independently; thus, both kinds of separation occur equally and all four type of gametes (RY, ry, Ry and rY) are produced in equal proportions. When these four types of gametes are combined to produce the F_2 generation, the progeny consists of 9/16 round and yellow, 3/16 wrinkled and yellow, 3/16 round and green, and 1/16 wrinkled and green. Thus the phenotypic dihybrid ratio is 9:3:3:1.

Dihybrid Test Cross

For working out crosses involving more than one character instead of Punnett square, a branch diagram or forked line method can be used. This is worked out by breaking the cross down into single-locus crosses and using the multiplication rule to determine the proportions of combinations of characteristics.

In a dihybrid test cross, the F_1 plants involved are heterozygous round and yellow (RrYy) and homozygous wrinkled and green (rryy). The cross is treated as a series of single-locus crosses. The cross Rr × rr yields 1/2 round (Rr) progeny and 1/2 wrinkled progeny (rr). The cross Yy × yy yields 1/2 yellow (Yy) progeny and 1/2 green (yy) progeny. Using the multiplication rule, the proportion of round and yellow progeny will be 1/2 × 1/2 = 1/4 (Figure 3.4).

Four combinations of traits with the following proportions appear in the offspring:

RrYy, round yellow	1/4
Rryy, round green	1/4
rrYy, wrinkled yellow	1/4
rryy, wrinkled green	1/4

The phenotypic ratio is $1:1:1:1$.

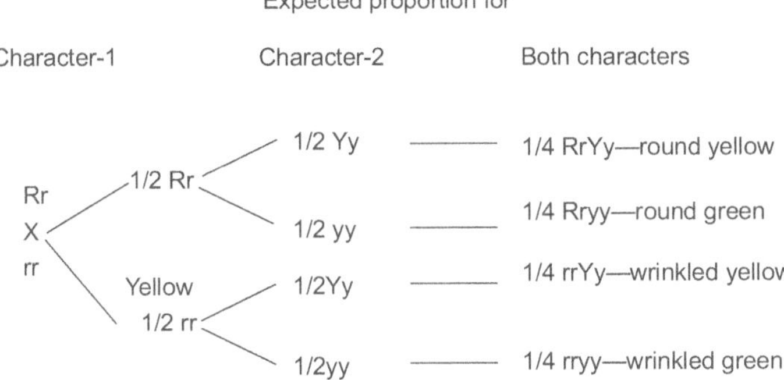

Figure 3.4   Dihybrid test cross (Branch diagram)

TRIHYBRID CROSS

The branch diagram can be applied to crosses including three characteristics, called trihybrid cross (Figure 3.5).

In one trihybrid cross, Mendel crossed a pure-breeding variety that possessed round seeds, yellow cotyledons and grey seed coats (RRYYGG) with another pure-breeding variety that possessed wrinkled seeds, green cotyledons and white seed coats(rryygg). The branch diagram shows that the expected phenotypic ratio in the F_2 is $27:9:9:9:3:3:3:1$.

1.	Round yellow grey	RYG	27/64
2.	Round yellow white	RYg	9/64

3.	Round green grey	RyG	9/64
4.	Round green white	Ryg	3/64
5.	Wrinkled yellow grey	rYG	9/64
6.	Wrinkled yellow white	rYg	3/64
7.	Wrinkled green grey	ryG	3/64
8.	Wrinkled green white	ryg	1/64

In monohybrid crosses, three genotypes are produced (RR, Rr, and rr) in F_2. In dihybrid crosses, 9 genotypes (3 genotypes for the first locus × 3 genotypes for the second locus = 9) are produced in the F_2. There are 3 possible genotypes at each locus when there are two alternative alleles. So the number of genotypes produced in the F_2 of a cross between individuals heterozygous for n loci will be 3^n.

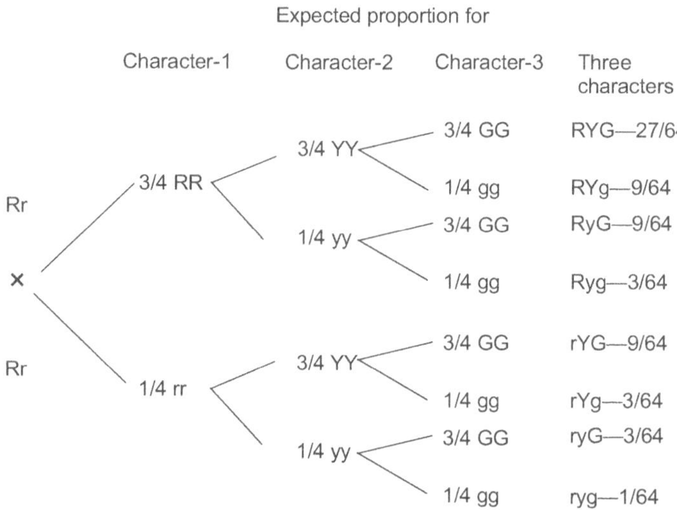

Figure 3.5   Trihybrid cross (Branch diagram)

Summary

- ✿ The basic principles of heredity were first proposed by Mendel.
- ✿ He conducted monohybrid and dihybrid crosses with *Pisum sativum*.
- ✿ The law of segregation states that each organism possesses two alleles for a trait, which segregate during gamete formation.
- ✿ The ratios of the genetic crosses can be predicted in Punnett square.
- ✿ The principle of independent assortment states that genes coding for different traits separate and assort independently in gamete formation.
- ✿ Trihybrid crosses can be represented by forked line method.

REVIEW QUESTIONS

1. Bring out the relation between the terms gene, alleles, genotype and phenotype.

2. Explain precisely the rules of dominance and segregation.

3. Mention the significance of a test cross.

4. What is the principle of independent assortment?

5. Present a multiple cross in a branch diagram.

4

INHERITANCE IN HUMANS AND HUMAN PEDIGREES

Objectives

- ✿ To learn the human Mendelian traits
- ✿ To understand the variable expression of genes
- ✿ To study the construction of human pedigrees
- ✿ To analyse the features of different kinds of pedigrees and the inheritance pattern

Key Terms

free ear lobe	tongue rolling	widow's peak
Mendelian traits	mid-digital hair	hitchhiker's thumb
bent little finger	thumb crossing	PTC tasting
pleiotropy	penetrance	polydactyly
modifiers	suppressor	phenocopy
phocomelia	pedigree analysis	achondroplasia
progeria	haemophilia	hairy pinna
colour blindness	hypophosphataemia	dimpled chin
sex-linked inheritance	Y-linked inheritance	cystic fibrosis
incomplete dominance	variable expressivity	
osteogenesis imperfecta	sickle-cell disease	
Duchenne muscular dystrophy		
autosomal recessive inheritance		
autosomal dominant inheritance		

INTRODUCTION

In all the living organisms phenotypic characters are expressed externally but are determined by the genes residing in the chromosomes. This was first recognized by Gregor Mendel. The characters controlled by genes are termed Mendelian traits. These traits are often determined by a pair of genes with alternate expression. The genes reside in the somatic chromosomes namely autosomes or in the sex chromosomes (XX or XY) and are called as autosomal traits or sex-linked traits respectively. Some traits are controlled by more than two alleles and are called multiple allelic traits.

MENDELIAN TRAITS

Many human traits follow Mendelian inheritance predictions. Some of the human traits are listed below.

1. **Eye colour**   If the eyes are brown, it is due to the presence of one dominant gene. Blue, grey and green eyes are recessive traits.

2. **Tongue rolling**   The ability to curl the tongue upwards from the sides is a dominant trait. The persons who cannot roll the tongue possess recessive genes.

3. **Free ear lobe**   Ear lobes that hang free from the ear are dominant over attached ear lobes that are attached directly to the side of the head.

4. **Widow's peak**   A distinctive downward point in the hairline is known as the Widow's peak. This is a dominant trait. If the hairline is straight then recessive genes are present.

5. **Hitchhiker's thumb**   The ability to bend the thumb backward (at least 45°) is caused by a dominant allele.

6. **Mid-digital hair**   The presence of hair on fingers is a dominant trait. Hair may not be present on all fingers,

but if hair is seen even in one finger, then this is dominant phenotype.

7. **Bent little finger**   A dominant gene causes the last joint of the little finger to bend inward towards the fourth finger. The straight little finger is a recessive trait.

8. **Thumb crossing**   Without thinking about it, hands are clasped together. If left thumb crosses over right then it is a dominant trait. On the other hand right over left is a recessive trait.

9. **Dimpled chin**   If dimple is noted in chin it is a dominant trait. If chin is not dimpled then it is recessive.

10. **PTC tasting**   Persons who can taste phenyl thiocarbamide (PTC) bitter are termed as tasters and this trait is controlled by a dominant gene. Non-tasters do not taste it and are recessive.

VARIATION IN GENE EXPRESSION

There are several important points to be analysed regarding the expression of a gene which was not shown in typical Mendelian inheritance. Though the gene responsible for a Mendelian trait controls a single phenotype, often genes tend to show many effects and are said to be pleiotropic (multiple effect). In the common example of sickle-cell anaemia, due to the sickle shape of the red blood cells, the blood vessels become clogged. There are also other symptoms like severe physical weakness, abnormal spleen, circulatory disturbances and brain damage. Such a group of symptoms characterizing a particular condition is termed as syndrome. Thus the sickle-cell gene is responsible for the appearance of several changes in the normal phenotype. The term pleiotropy refers to the collection of effects associated with a specific gene.

The heterozygotes for sickle-cell anaemia carry one normal allele and another sickle-cell allele ($Hb^A Hb^S$). They are carriers.

Though they are healthy, they are not identical to the normal persons who carry two normal alleles ($Hb^A Hb^A$). The heterozygous condition is termed as sickle-cell trait. This difference from the normal type is evident in blood test. In heterozygotes, some cells become abnormally shaped under low oxygen tension. The terms dominant and recessive cannot be applied to this pair of alleles Hb^A Hb^S since both kinds of haemoglobins are present. So it is evident that both alleles express themselves in the heterozygote. Such genes are said to be codominant. The term incomplete dominance is often used to describe conditions where the effects of one of the alleles in the heterozygotes are more pronounced than the other.

The expression of a gene may also vary from person to person. That is, the expression is not constant and varies from one individual to the other. This is said to be **variable expressivity**. In the disease osteogenesis imperfecta, the gene (O) responsible for the disease is dominant while its normal allele is recessive (o). In the presence of the gene (O) a syndrome is established. There is severe fragility of bones, weakness of ligaments and tendons, deafness and a blue colouration of the eye. However when persons afflicted with osteogenesis imperfecta were studied, it has been shown that affected persons do not exhibit all the symptoms. For example, a parent may exhibit all the symptoms while the children show only blue eye colouration. It is clear that the gene does not express itself to the same extent or in the same way in every person who carries it.

Another feature of a gene is **penetrance**. Penetrance is defined as the percentage of individuals having a particular genotype that expressed the expected phenotype. At times, certain genes, a dominant one or a recessive one in the homozygous condition, may remain unexpressed when present in the genotype. The gene for osteogenesis imperfecta, when present, does not bring out a detectable effect in all individuals. Nine out of ten persons carrying the dominant gene will show one or more symptoms, while the remaining one will be normal. The gene has reduced penetrance. That is, the gene has a penetrance of 90% expressing itself in some way. On the other hand, the genes for ABO blood types are 100% penetrant, that is, when they are present, they will always be present 100%. The gene for osteogenesis imperfecta is pleiotropic showing reduced penetrance and variable expressivity.

Another example is polydactyly, that is, extra digits in hands and foot which show incomplete penetrance. The trait is caused by a dominant allele. In some cases the parents possess the allele but do not show polydactyly: but their children inherit polydactyly. In this case the gene for polydactyly is not fully penetrant. For example, if 42 people having the polydactylous gene are examined and only 38 of them are polydactylous, the penetrance would be $38/42 = 0.9$ (90%). Expressivity is a related concept. It is the degree of expression. In addition to incomplete penetrance, polydactyly shows variable expression. Some polydactylous persons possess extra fingers and toes that are fully functional, whereas others possess only a small tag of extra skin.

The expression of any one gene may be greatly influenced by other genes present in the genotype. Generally no genes act independently of other genes. The presence of one gene may completely prevent the expression of another. For example eye colour involves not just one pair of alleles but the interaction of many. At least three different genes have been identified which reduce the amount of melanin pigment in both skin and eyes. The third gene causes a lack of pigment only in the eye.

Certain genes, known as **modifiers**, have a slight effect on the expression of some other gene altering the expression in a quantitative way. An abnormal effect may be completely suppressed by the presence of some other gene. Such genes which prevent the expression of some other genes are called **suppressors**.

Interaction of genes is also influenced by environment. An individual whose phenotype has been environmentally altered so that it produces a phenotype that is the same as the phenotype produced by a genotype is called **phenocopy**. The familiar example is that caused by the drug thalidomide, an ingredient in certain sleeping pills. The thalidomide tragedy became evident in Europe with an increase in the birth of babies without limbs called **phocomelia** following the consumption of this drug by pregnant women. The victims phenotypically resemble those individuals who have reduced limbs due to a rare hereditary disorder produced by a dominant gene. However the thalidomide victims possess a normal genotype for limb development. Though they are limbless these individuals will transmit only the normal genes. In the study of human genetics, gene interactions must be considered and the influence of environment in gene expression is never to be ignored.

HUMAN PEDIGREES

The study of inherited Mendelian traits in humans must rely on observations made while working with individual families. Classical cross-fertilization breeding experiments as performed by Mendel are not allowed in humans. Human geneticists are not allowed to selectively breed for the traits they wish to study. One of most powerful tools in human genetic studies is pedigree analysis. When human geneticists first began to publish family studies, they used a variety of symbols and conventions. Now there are agreed-upon standards for the construction of pedigrees.

Pedigree analysis is one of the best methods of analysing inheritance. It determines the mode of inheritance of a trait.

Pedigree is a chart or diagram representing the ancestral history of an individual (Figure 4.1). It is a short-hand or graphic representation of the details of a family. Individuals are characterized according to their sex, their generation, and their biological relationship to each other.

Typically, a three-generation pedigree is obtained, beginning with the patient. For example, if an infant is being evaluated, the pedigree should include siblings, parents, aunts and uncles, nieces and nephews and grandparents. When evaluating an adult, it is often helpful to expand the pedigree to include children and grandchildren. If a pattern of illness emerges, it is important to extend the family history back to as many generations as possible to include any additional affected relatives.

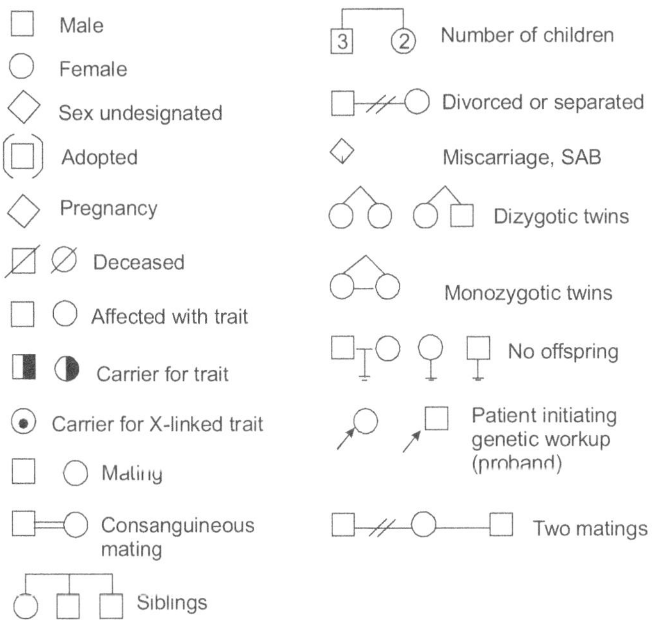

Figure 4.1   Symbols applied in a pedigree
(*Source*: www.uci.edu)

The factual and health information to include in a pedigree are the following.

- ☼   Age/birth date
- ☼   Age of death
- ☼   Cause of death
- ☼   Pregnancy complications (e.g. miscarriage, stillbirth, pregnancy termination)
- ☼   Infertility vs. number of children by choice
- ☼   Health information (height, weight, etc.)
- ☼   Affected/unaffected status (defined by shading of symbols in key/legend)
- ☼   Ethnic background
- ☼   Consanguinity
- ☼   Date when pedigree was drawn
- ☼   Name of the person who provided the data

Males are always represented by square symbols, females with circular symbols. A line drawn between a square and a circle represents a mating of that male and female. Two lines drawn between a square and a circle indicate a consanguineous mating, the two individuals are related, usually second cousins or closer relatives. When possible, the square should be placed on the left and the circle on the right of the mating line. Generations are connected by a vertical line extending down from the mating line to the next generation. Children of a mating are connected to a horizontal line, called the sib-ship line, by short vertical lines. The children of a sib-ship are always listed in order of birth, the oldest being on the left. Normal individuals are represented by an open square or circle, depending upon the gender, and affected individuals by a solid square or circle. Each generation is numbered to the left of the sib-ship line with Roman numerals. Individuals in each generation are numbered sequentially, beginning on the left, with Arabic numerals. For example, the third individual in the second generation would be identified as individual II-3.

INHERITANCE PATTERNS

Autosomal Dominant Inheritance

The pattern of autosomal dominant inheritance is perhaps the easiest type of Mendelian inheritance to recognize in a pedigree. One allele responsible for a trait is all that is required for the expression of the phenotype. The features for this pattern are given below. With the understanding that almost all affected individuals are heterozygotes, and that in most matings involving a person with an autosomal dominant trait the other partner will be homozygous normal, there are four hallmarks of autosomal dominant inheritance.

1. Generally every affected individual has an affected biological parent.

2. The transmission of the trait is from generation to generation without skipping generation.

3. Males and females have an equally likely chance of inheriting the mutant allele and being affected.

4. In the mating between an affected heterozygote and a normal, each child has 50% chance to inherit the abnormal allele and be affected and a 50% chance to inherit the normal allele.

The family represented in Figure 4.2 is a good example of how autosomal dominant diseases appear in a pedigree. Each of the four hallmarks of autosomal dominant inheritance is fulfilled. About 50% of the offspring of an affected individual are affected (recurrence risk is 50%). Normal offspring II-3 of affected individuals have all normal offspring.

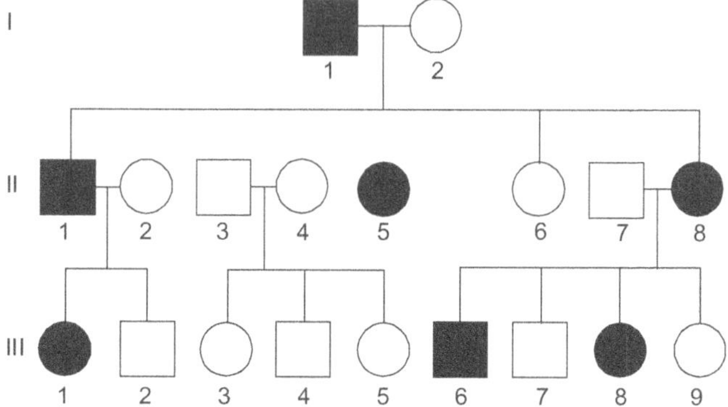

Figure 4.2　Sample pedigree for autosomal dominant inheritance (*Source*: www.uic.edu)

Examples

1. **Achondroplasia (dwarfism)**　AA = homozygous dominant is lethal (fatal spontaneous abortion of foetus), Aa = dwarfism, aa = no dwarfism. 99.96% of all people in the world are homozygous recessive (aa).

2. **Polydactyly (extra fingers or toes)**　PP or Pp = extra digits, pp = 5 digits. 98% of all people in the world are homozygous recessive (pp).

3. **Progeria (very premature aging)**　Spontaneous mutation of one gene creates a dominant mutation that rapidly accelerates aging.

Autosomal Recessive Inheritance

The first, and most important thing to remember about autosomal recessive inheritance is that most, if not all, affected individuals have parents with normal phenotypes.

1. Males and females are equally likely to be affected.

2. On average, the recurrence risk to the unborn sibling of an affected individual is 25%.

3. The trait is characteristically found in siblings, not parents of affected or the offspring of affected.

4. Parents of affected children may be related and a consanguineous mating is involved.

5. The trait often skips generation.

6. If both parents are affected then all the offspring are affected.

The pedigree shown in Figure 4.3 illustrates four hallmarks of autosomal recessive inheritance. I-1 and I-2 are unrelated, yet they produced an affected offspring (affected offspring have normal parents). By chance, they both must have been carriers.

When consanguinity is involved, i.e., matings between related individuals, in the production of an affected child, the assignment of probabilities changes, especially in the rare autosomal recessive diseases. Consanguinity introduces the possibility of one founding parent being a carrier, with the recessive allele being passed through carrier offspring and produces an affected homozygous offspring some generations later. When an affected child is produced as the result of a consanguineous mating, those individuals in the direct line of descent are most probably carriers and those from outside the family are most probably normal homozygotes. V-1 is affected with an autosomal recessive disease. Her parents are second cousins. IV-1 and IV-2 must both be carriers since they produced an affected child. (The child must have received a recessive allele from each of her parents.) III-2 is an obligate carrier. Her father was affected, and hence, a homozygote for the recessive allele. III-5 must also be heterozygotes since IV-2 had to get her recessive allele from one of her parents, and the chance of III-6 being a carrier is less than 1 in 50. I-1 and I-2 must have been carriers since they produced an affected offspring, II-1.

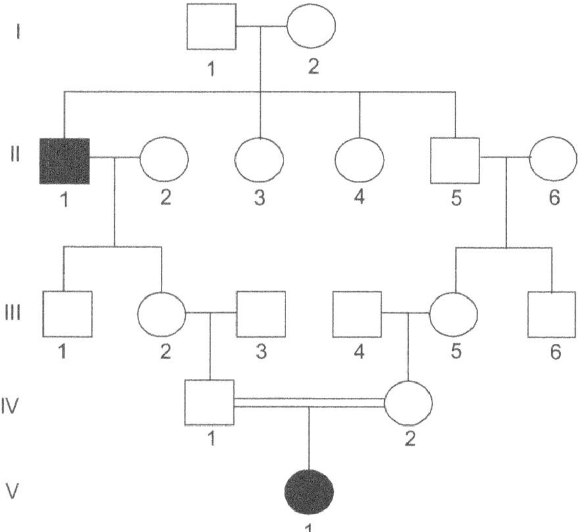

Figure 4.3   Pedigree for autosomal recessive inheritance (*Source*:www.nature.com)

Examples

1. **Cystic fibrosis**   Homozygous recessives (cc) have cystic fibrosis—body cannot make needed chloride channel, high concentrations of extracellular chloride causes mucus to build up, infections and pneumonia.

2. **Sickle-cell disease**   The most common inherited disease of African–Americans (1:400 affected). Homozygous recessives (ss) make abnormal form of haemoglobin that deforms red blood cells (sickle-shaped) and causes a cascade of symptoms like clogging of blood vessels, organ damage and kidney failure.

Sex-linked Inheritance

If the gene for a particular trait or disease lies on the X chromosome, the disease is said to be X-linked. The inheritance pattern for X-linked inheritance differs from autosomal inheritance only because the X chromosome has no homologous chromosome in the male, the male has an X and a Y chromosome. Very few genes have been discovered on the Y chromosome.

The inheritance pattern follows the pattern of segregation of the X and Y chromosomes in meiosis and fertilization. A male child always gets his X from his mother's two X's and his Y chromosome from his father. X-linked genes are never passed from father to son. A female child always gets the father's X chromosome and one of the two X's of the mother. An affected female must have an affected father. Males are always hemizygous for X-linked traits, that is, they can never be heterozygotes nor homozygotes. They are never carriers. A single dose of a mutant allele will produce a mutant phenotype in the male, whether the mutation is dominant or recessive. On the other hand, females must be either homozygous for the normal allele, heterozygous, or homozygous for the mutant allele, just as they are for autosomal loci.

X-linked dominant inheritance   When an X-linked gene is said to express dominant inheritance, it means that a single dose of the mutant allele will affect the phenotype of the female. A recessive X-linked gene requires two doses of the mutant allele to affect the female phenotype (Figure 4.4). The following are the hallmarks of X-linked dominant inheritance:

1. The trait is never passed from father to son.

2. All daughters of an affected male and a normal female are affected. All sons of an affected male and a normal female are normal.

3. Mating of affected females and normal males produce 1/2 the sons affected and 1/2 the daughters affected.

4. In the general population, females are more likely to be affected than males, even if the disease is not lethal in males.

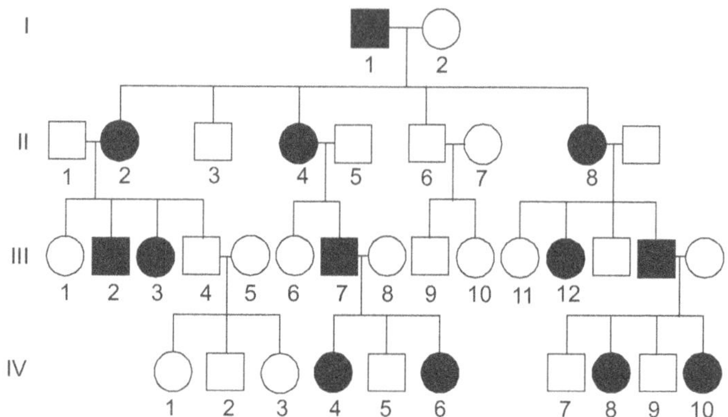

Figure 4.4   Pedigree for X-linked dominant inheritance
(*Source*: www.nature.com)

The following Punnett squares (Table 4.5a and b) explain
the first three hallmarks of X-linked dominant inheritance.
X represents the X chromosome with the normal allele,
X^A represents the X chromosome with the mutant dominant
allele, and Y represents the Y chromosome. Note that the
affected father never passes the trait to his sons but passes it
to all of his daughters, since the heterozygote is affected for
dominant traits. On the other hand, an affected female passes
the disease to half of her daughters and half of her sons.

		Father's gametes				Father's gametes	
		X^A	Y			X	Y
Mother's gametes	X	XX^A	XY	**Mother's gametes**	X^A	$X^A X$	$X^A Y$
	X	XX^A	XY		X	XX	XY
		(a)				(b)	

Figure 4.5   (a) Mating between affected father and
normal mother   (b) Mating between affected
mother and normal father

Males are usually more severely affected than females because in each affected female there is one normal allele producing a normal gene product and one mutant allele producing the non-functioning product, while in each affected male there is only the mutant allele with its non-functioning product and the Y chromosome, no normal gene product at all. Affected females are more prevalent in the general population because the female has two X chromosomes, either of which could carry the mutant allele, while the male has only one X chromosome as a target for the mutant allele. When the disease is no more deleterious in males than it is in females, females are about twice as likely to be affected as males. As shown in Figure 4.5a and b, X-linked dominant inheritance has a unique heritability pattern.

The key for determining, whether a dominant trait is X-linked or autosomal is to look at the offspring of the mating of an affected male and a normal female. If the affected male has an affected son, then the disease is not X-linked. All of his daughters must also be affected if the disease is X-linked.

X-linked dominant inheritance is relatively rare. One example of X-linked dominant inheritance is hypo-phosphataemia. This disease is characterized by abnormally low levels of phosphorus in the blood due to defective reabsorption of phosphate in the kidneys. Calcium and vitamin D metabolism are also abnormal resulting in softening of bones and rickets and this condition is called vitamin D-resistant rickets.

X-linked recessive inheritance   Everyone has heard of some X-linked recessive disease even though they are, in general, rare. Haemophilia, Duchenne muscular dystrophy, Lesch–Nyhan syndrome are relatively rare in most populations, but because of advances in molecular genetics they received attention in the media. More common traits, such as glucose 6-phosphate dehydrogenase deficiency or colour blindness, may occur frequently enough in some populations to produce a few affected females. However, their effect on individuals is rarely life-threatening and medical intervention is not needed.

The features of X-linked recessive inheritance (Figure 4.6) are:

1. The trait is never passed from father to son.

2. All daughters of an affected male and a normal female are affected. All sons of an affected male and a normal female are normal.

3. Mating of affected females and normal males produce 50% of sons to be affected and 50% of the daughters to be affected.

4. In the general population, females are more likely to be affected than males, even if the disease is not lethal in males.

5. The trait does not skip generations.

Normal mother's gametes	Affected father's genotype X^A	Affected father's genotype Y	Carrier mother's genotype	Normal father's genotype X	Normal father's genotype Y
X	XX^A	XY	X^A	X^AX	X^AY
X	XX^A	XY	X	XX	XY
All daughters carriers, all sons normal			Half of sons affected, half of daughters affected		

Figure 4.6   X-linked recessive inheritance

In Figure 4.7, II-2 and II-5 are both carriers, their father was affected and passed on his only X chromosome to his daughters. II-3 cannot be a carrier for two reasons. First, males

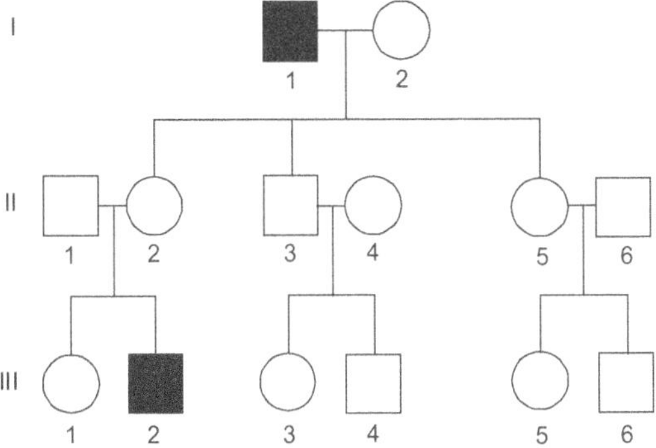

Figure 4.7　Pedigree for X-linked recessive inheritance
(*Source*: www.nature.com)

are either affected or normal, never carriers. Second, he did not inherit his father's X chromosome. He inherited his father's Y chromosome. III-3 could not have been a carrier since neither her father nor her mother had the mutant gene.

Examples

1. **Colour blindness**　This term is usually applied to those who cannot perceive colours. Most commonly observed type is red/green colour blindness. This is a standard example of sex-linked recessive inheritance in man and is harmless.

2. **Haemophilia**　It is a sex-linked recessive disorder. In this condition, absence or deficiency of a specific protein clotting factor greatly slows down blood clotting.

3. **Duchenne Muscular dystrophy (DMD)**　DMD is an X-linked recessive disease occurring in about 1 in 3,500 newborn males. The normal muscle protein dystrophin is absent, causing muscle weakness. Usually the onset occurs before age 6 years, and the victim is chair-ridden by age 12 and dead by age 20.

Y-linked Inheritance

Very few genes have been detected on the Y chromosome. In humans Y is less than half the length of the X chromosome. The features are:

1. Trait expression and transmission is only in males, the individuals with the Y chromosome.

2. If a male has a trait, so should his father and paternal grandfather as well as his sons and their sons. It follows the inheritance of the Y chromosome.

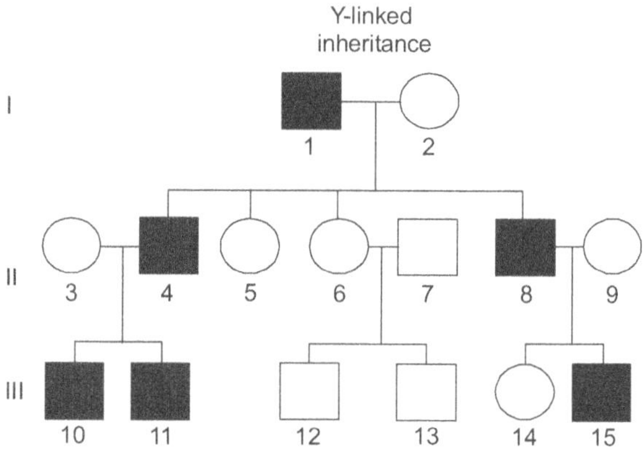

Figure 4.8   Pedigree for Y-linked inheritance
(*Source.* www.nature.com)

As shown in Figure 4.8, pedigree all the male descendants of I-1 are affected. The sons III-12 and13 of II-7 are not affected since they carry the Y of their unaffected father II-7. One example is the hairy pinna, where there is excessive growth of hair in the external ear.

Summary

- The traits controlled by genes are termed Mendelian traits.
- Genes may either reside in the autosomes or sex chromosomes.
- There are a number of Mendelian traits in humans.
- Genes often have variable expressions indicated by penetrance, pleiotropism, incomplete dominance and phenocopy.
- Inheritance of a genetic trait in humans can be studied by analysing pedigrees.
- Pedigree is a digrammatic representation of a family history.
- A number of specific symbols are used in constructing a pedigree.
- The kinds of inheritance patterns are autosomal dominant and recessive inheritance, X-linked dominant and recessive inheritance and Y-linked inheritance.
- Each pattern has unique features that could be used in identifying the type of inheritance of a trait.

REVIEW QUESTIONS

1. Define Mendelian traits and give examples.

2. With reference to variable gene expression, explain penetrance, pleiotropy and phenocopy, giving examples.

3. Describe the symbols commonly used while constructing a human pedigree.

4. Draw a typical pedigree showing all the features for

     i. autosomal dominance

     ii. sex-linked dominance

5. How will you distinguish the pedigree of sex-linked recessive inheritance from Y-linked inheritance?

6. Justify the statement that the terms homozygous and heterozygous are not applicable to X-linked genes in males.

5

GENE INTERACTION AND MODIFIED MENDELIAN RATIOS

Objectives

- ☼ To view the kinds of gene interactions
- ☼ To analyse the modifications of Mendelian ratios

Key Terms

gene interaction	non-allelic	epistasis
hypostasis	anthocyanin	lethal gene
creeper chicken	yellow lethal	polygenes
quantitative inheritance	mulatto	
complementary gene	supplementary gene	

INTRODUCTION

In the previous chapter, emphasis was placed on the fact that the genes studied by Mendel were segregating independently of each other and the genes also have been functioning independently of one another. But generally do not occur and function separately and independently in cells. The genes are all located in the nuclei of the cells. Thus, it is natural that the expression of an allele of one gene will sometimes alter the expression of one or more of the alleles of a second (non-allelic) gene. So most of the times the Mendelian monohybrid ratio of 3:1 and dihybrid ratio of 9:3:3:1 are not observed. Thus the Mendelian ratios are modified by gene interactions.

The following examples are some of the known gene interactions.

GENE INTERACTION THAT PRODUCES NOVEL PHENOTYPES (COMB PATTERN IN FOWLS)

A classic example of gene interaction based on the results of crosses between different breeds of chicks was reported by William Bateson and R.C. Punnett. They proved in this example that two gene pairs, which are present on separate loci but influence the same trait, interact to produce some totally new trait or phenotype that are not represented in the parents.

Domestic breeds of chickens have different comb shapes. Wyandottes have a characteristic type of comb called rose whereas brahmas have a pea comb and leghorns have single combs. When wyandottes and Brahmas were crossed, all F_1 chickens had "**walnut**" combs, a phenotype not expressed in either parent. When the F_1 chickens were mated among themselves, in F_2 the familiar ratio of $9 : 3 : 3 : 1$ was produced but of the four kinds of phenotypes, two types were different from their parents.

About 9/16 of the F_2 hybrids were walnut, 3/16 were rose, 3/16 were pea and 1/16 had single combs. Neither single comb nor walnut was expressed in the original parental lines. This is the result of interaction of genes (Figure 5.1). A gene for rose comb and a gene for pea comb would interact and produce walnut as in the F_1 generation.

Analysis of the F_2 results indicated that 9/16, with the two dominant genes (R, P) was walnut like F_1 chickens; 1/16 representing full recessive combinations (rrpp) was characterized by single comb. The rose (Rpp) and pea comb (rrP) were 3/16.

It was determined that the homozygous genotype of the rose-combed parent (Wyandotte) was RRpp and that of the pea-combed parent (Brahma) was rrpp. Though the usual $9:3:3:1$ ratio was obtained, the result from this cross was unusual in two important respects, (i) The F_1 progeny differed from the parents, that is none was rose or pea but all were walnut. (ii) Two phenotypes (walnut and single) not expressed in the original parents appeared in the F_2.

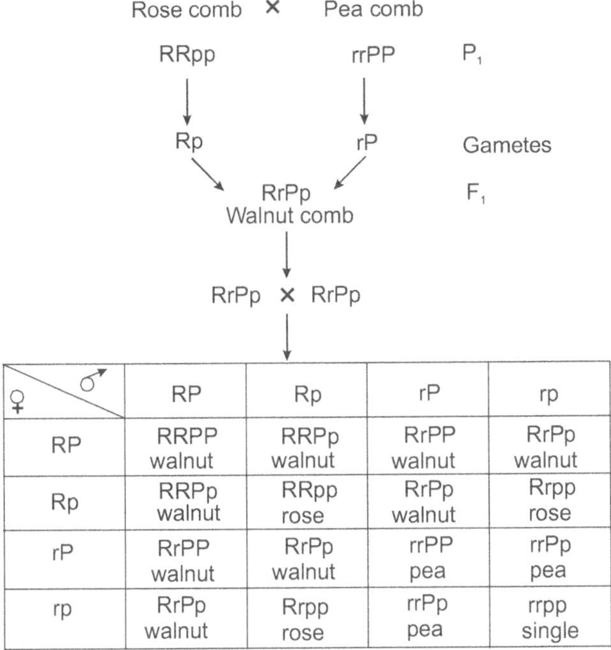

♀ \ ♂	RP	Rp	rP	rp
RP	RRPP walnut	RRPp walnut	RrPP walnut	RrPp walnut
Rp	RRPp walnut	RRpp rose	RrPp walnut	Rrpp rose
rP	RrPP walnut	RrPp walnut	rrPP pea	rrPp pea
rp	RrPp walnut	Rrpp rose	rrPp pea	rrpp single

F_2 generation — 9 Walnut : 3 Rose : 3 Pea : 1 Single (9:3:3:1)

Figure 5.1   Inheritance of comb pattern in fowls

Genes R and P were non-allelic, but each was dominant over its allele (R over r and P over p). When R and P were together, as in F_1 (RrPp), the two different products interacted to produce a walnut comb. The two non-allelic genes R and P acted independently in different ways to produce new combinations.

EPISTASIS

Epistatic interaction means that one gene masks or changes the effect of another gene. In other words, it does not allow the other gene to express itself fully. Epistasis may be of the following types:

 i. Due to a recessive allele, i.e., A masks the effect of B
 ii. Due to a dominant allele, i.e., A masks the effect of B

In some cases, this may be in only one direction meaning that either A masks the effect of B or vice versa. In other cases, it can be in both directions also, which means that both genes are mutually epistatic to each other. In epistasis, the gene that does the masking is called the **epistatic gene**, the gene whose effect is masked is a **hypostatic gene**.

An example of epistasis is the body colour in mice.

Body Colour in Mice

In mice, the wild body colour is known as agouti, characterized by banding of individual hairs. The agouti colour is controlled by a gene A, which is hypostatic to the recessive allele c. The dominant allele C in the absence of A gives coloured mice. Moreover, in the presence of dominant allele C, A gives rise to agouti. Therefore, CCaa will be coloured and ccAA will be albino.

When coloured mice CCaa are crossed with albino ccAA, agouti mice CcAa appear in F_1. The details of the phenotypes and genotypes obtained in F_2 are given in Table 5.1. As can be realized, cc masks the effect of AA in this example and is therefore epistatic. Consequently, ccAA is albino. However, ccaa is also albino, but not due to epistasis but due to the absence of both the dominant alleles.

Table 5.1   Genotypic ratio and phenotypic ratio obtained in F_2

	CA	Ca	cA	ca
Genotypic ratio	9 :	3 :	3 :	1
Phenotypic ratio	agouti	coloured	albino	
	9 :	3 :	4	

COMPLEMENTARY INTERACTION

The complementary genes are two pairs of non-allelic dominant genes (present on separate gene loci), which interact to produce only one phenotype trait, but neither of them when present alone in the absence of other produces the phenotypic trait. It means that for the development of the dominant character in question, both the complementary genes must be represented at least by their one dominant allele. Absence of even one of the two genes produces recessive phenotype.

Flower colour in sweet pea and deaf-mutism in man are examples of complementary interaction.

Flower Colour in Sweet Pea

In sweet pea (*Lathyrus odoratus*), the purple colour of flowers is dependent on two non-allelic complementary genes C and P. Gene C produces an enzyme that catalyses the formation of colourless chromogen and formation of anthocyanin pigment. Gene P controls the production of an enzyme which catalyses the transformation of this chromogen into anthocyanin. These genes are complementary to each other. It means the pigment anthocyanin is produced by two biochemical reactions and the end product of the first reaction forms the substrate for the other.

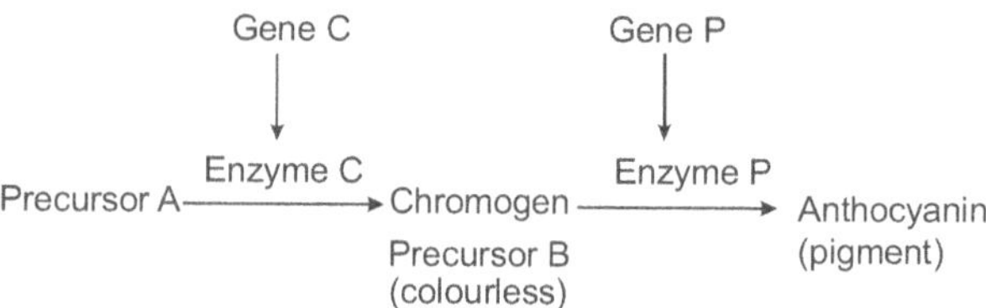

If a plant possesses dominant genes C and P, it produces purple flowers. But if a plant has a genotype CCpp, it produces raw material but is unable to convert it into anthocyanin. Therefore, it produces white flowers. Similarly, if it possesses dominant gene P, but no dominant C [ccPP], it produces white flowers because gene P can convert colourless chromogen into anthocyanin but cannot form chromogen.

The dominant allele or alleles (CC or Cc) of gene are responsible for the presence of chromogen, while the homozygous recessive alleles (cc) of this gene are responsible for the absence of chromogen. Likewise, the dominant alleles of gene P in homozygous (PP) or heterozygous (Pp) conditions caused the production of an enzyme which is necessary for colour production from chromogen, while homozygous recessive (pp) condition does not produce any such enzyme. The apperance of 9 coloured and 7 white flowers (9 : 7) is the modified Mendelian dihybrid ratio.

Bateson artificially crossed two strains of *Lathyrus*, the progeny had purple flowers. These purple individuals of F_1 when self-pollinated produced purple and white in the ratio of 9 : 7. Bateson assumed that 9 : 7 is the mere modification of Mendelian F_2 ratio 9 : 3 : 3 : 1. It means that the two white strains of *Lathyrus* actually have the genotype ccPP and CCpp. The result of the above cross is presented in Figure 5.2.

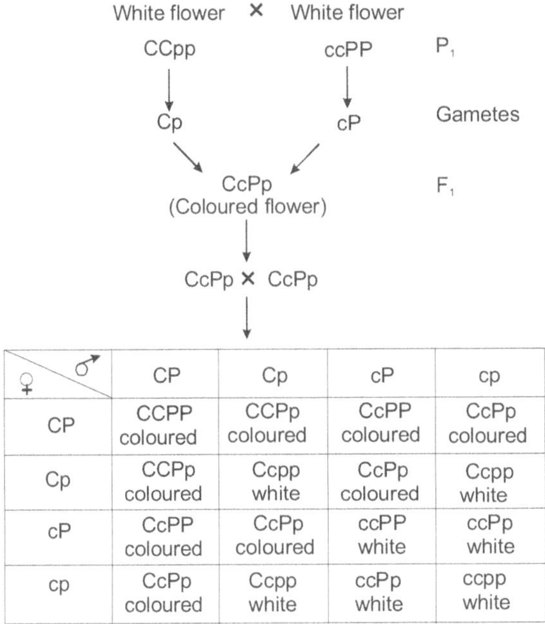

♀ \ ♂	CP	Cp	cP	cp
CP	CCPP coloured	CCPp coloured	CcPP coloured	CcPp coloured
Cp	CCPp coloured	Ccpp white	CcPp coloured	Ccpp white
cP	CcPP coloured	CcPp coloured	ccPP white	ccPp white
cp	CcPp coloured	Ccpp white	ccPp white	ccpp white

F$_2$ generation—9 coloured flower : 7 White flower (9 : 7)

Figure 5.2   Inheritance of flower colour in sweet pea

Deaf-Mutism in Man

In man, deaf-mutism is an example of complementary genes. The normal hearing and speech in man develops as a result of interaction between dominant genes A and B. A person homozygous for the recessive alleles of either of the two genes (AAbb or aaBB) is a deaf mute.

A cross between deaf mute aaBB and AAbb produces AaBb normal in F$_1$ generation. The cross between AaBb and AaBb results in 9 normal and 7 deaf mute in F$_2$ generation.

SUPPLEMENTARY INTERACTION

Supplementary genes are two independent pairs of dominant genes which interact in such a way that one dominant gene will produce its effect whether the other is present or not. The second dominant when added changes the expression of the first one but only in the presence of first one. Coat colour in rats is the result of supplementary gene interaction.

Coat Colour in Rats

In rats and guinea pigs, coat colour is governed by two dominant genes A and C, the agouti coloured guinea pigs have genotype CCAA. The black rat possesses gene for black colour (C) but not the gene (A) for agouti colour. If a gene for black colour is absent agouti is unable to express itself and rats with a genotype ccAA are albino. Here presence of gene C produced black colour and addition of gene A changes its expression to agouti colour. The result is 9 agouti : 3 black : 4 albino (Figure 5.3).

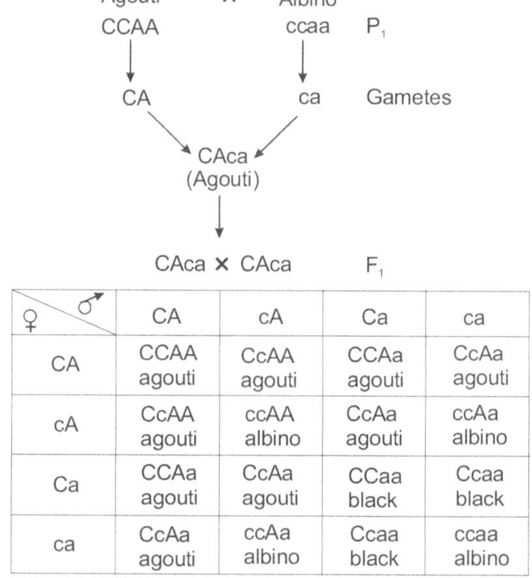

♀ \ ♂	CA	cA	Ca	ca
CA	CCAA agouti	CcAA agouti	CCAa agouti	CcAa agouti
cA	CcAA agouti	ccAA albino	CcAa agouti	ccAa albino
Ca	CCAa agouti	CcAa agouti	CCaa black	Ccaa black
ca	CcAa agouti	ccAa albino	Ccaa black	ccaa albino

Figure 5.3   Inheritance of coat colour in rats

LETHAL INTERACTION

Genes may affect viability as well as the phenotypic traits of an organism. Experiments have shown that animals carrying certain genes perform impaired biochemical as well as physical functions. Some genes do not affect the appearance of an organism but do influence viability. Some genes have such serious effects that the organism is unable to live. These are called lethal genes. If the lethal effect is dominant, all individuals carrying the genes will die and the gene will be lost. Some dominant lethals have delayed effect, thus the organism lives for some time. Recessive lethals carried in the heterozygous condition have no effect but may be expressed when a cross occurs between two carriers. Creeper chickens and yellow lethal in mice are examples of lethal interaction.

Creeper Chickens

The dominant gene *C* in chickens is responsible for developmental changes that result in aberrant forms called "Creepers" and the homozygous genotype CC is lethal. These birds have short, crooked legs and are of little value. When two creepers were crossed, a ratio of 2 creepers to 1 normal instead of 3 : 1 appeared. This is the characteristic ratio for lethal interactions. The creepers are heterozygous Cc and the missing 1/4 individuals were CC.

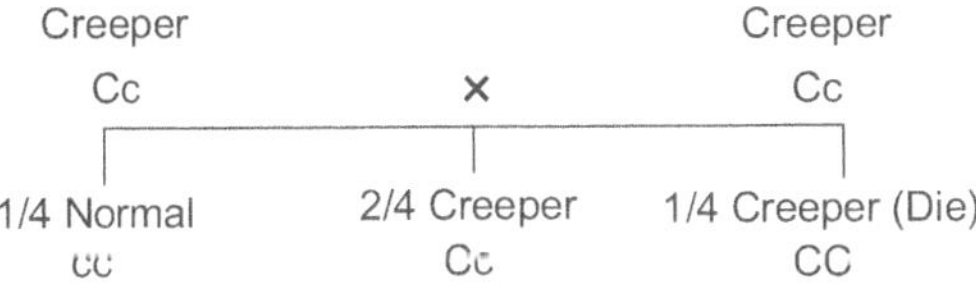

The creeper gene in homozygous condition produces severe malformations leading to death during incubation.

Yellow Lethal in Mice

Chenot found that the yellow mice never breed true. Whenever the yellow mice were crossed with yellow mice, always yellow and brown mice were obtained in the ratio of 2:1. It was concluded that yellow mice are heterozygous. The homozygous yellow (1/4) die in the embryonic condition. The gene *Y* has a multiple effect. (i) It controls yellow body colour and has a dominant affect. (ii) It affects viability and acts as a recessive lethal.

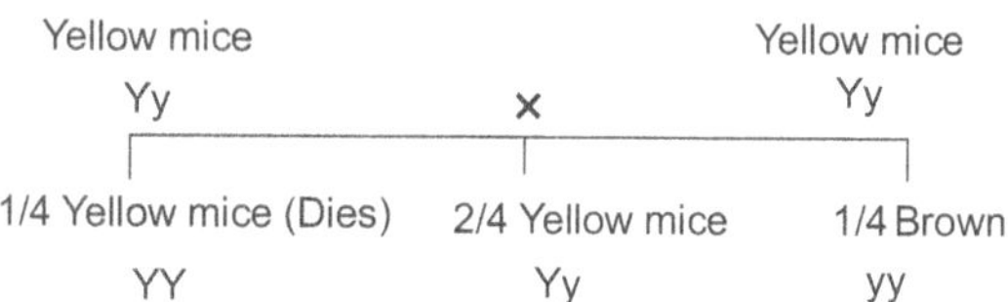

POLYGENIC INHERITANCE

The majority of genetic traits exhibit a minimum of two contrasting phenotypes. Such traits are designated as qualitative or discontinuous traits. For example, in garden pea plants the seed is either yellow or green and the plants are either tall or dwarf; in cattle, they may have horns or no horns; humans may have blood group A or B or AB or O ; *Drosophila* may have red eyes or white eyes.

But there are traits where a gradation in the trait is observed. For example, if a group of persons is classified according to their heights, a gradual gradation in the height can be easily noted. Such traits exhibit continuous phenotypic variation. F.C. Galton (1883) noted that many of these continuous variations are observed in human beings and are inherited. Characters like height, weight, skin colour or intelligence exhibit gradual differences. These characters are often determined by a number of genes and all of them have cumulative or additive effect. Each gene contributes a certain amount of effect. The degree of expression of a trait then depends upon the number of genes present. If the number of dominant genes are more, then the degree of expression of the character will be more. This kind of inheritance is termed as quantitative inheritance or polygenic inheritance. The genes involved in this kind of inheritance are known as polygenes. So polygenes are defined as two or more different pairs of genes which are non-allelic, having cumulative effect and are responsible for quantitative characters.

Characteristics of Polygenes

1. The effects of each contributing gene are cumulative or additive.

2. Each contributing gene produces an equal effect.

3. There is no dominance involved.

4. The polygenes have pleiotropic effects, that is, one gene may modify or suppress more than one phenotypic trait. A single gene may do only one thing chemically but may affect many characters.

5. The environmental conditions have considerable effect on the phenotypic expression of polygenes. In plants for example the height may be altered by soil, water, temperature, light or nutrition. Similarly identical twins with similar genotypes, if grown up in different environments, show different IQ. The skin colour of one person can be lighter or darker depending upon the amount of exposure to sunlight.

Some examples of polygenic inheritance are the following.

Kernel Colour in Wheat

To illustrate how multiple genes interact, Nilsson–Ehle studied kernel colour in wheat. He found that the intensity of red pigmentation was determined by three separate loci, each of which had two alleles. He crossed a variety of wheat that possessed white kernels with a variety that possessed purple kernels and obtained the following results.

Nilsson-Ehle interpreted the results as the result of aggregation of alleles at two loci. He proposed that there were two alleles at each locus: one that produced red pigment and another that produced no pigment. The alleles that encoded the pigment is designated as A^+ and B^+ and the alleles that encoded no pigment as A^- and B^- (Table 5.2). Each gene contributes equally to the pigmentation and the effects of the genes were additive. So the overall phenotype could be determined by adding the effects of all the genes.

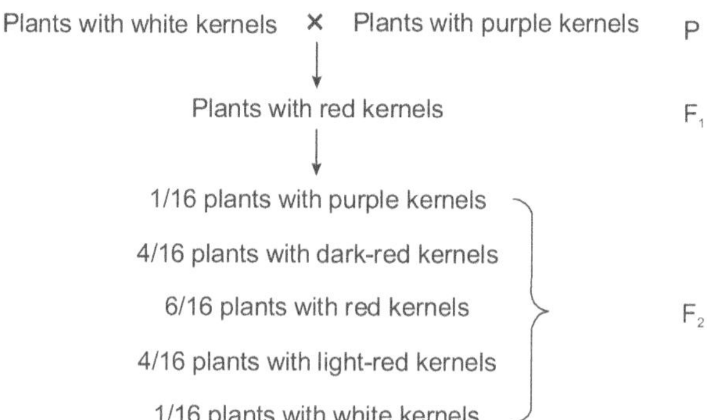

Table 5.2   The range of phenotypes

Genotype	Doses of pigment	Phenotype
$A^+ A^+ B^+ B^+$	4	Purple
$A^+ A^+ B^+ B^-$		
$A^+ A^- B^+ B^+$	3	Dark-red
$A^+ A^+ B^- B^-$		
$A^- A^- B^+ B^+$		
$A^+ A^- B^+ B^-$	2	Red
$A^+ A^- B^- B^-$		
$A^- A^- B^+ B^-$	1	Light-red
$A^- A^- B^- B^-$	0	White

It is observed that five phenotypes are possible when alleles at two loci influence the phenotypes and their effects are additive.

Skin Colour in Man

A classical example of polygenic inheritance was given by Davenport (1913) in Jamaica. The presence of melanin pigment in the skin of man determines the skin colour. The more the pigment the darker is the skin colour. The amount of melanin developing in an individual is determined by two pairs of genes. These genes are present at two different loci and each dominant gene is responsible for the synthesis of fixed amount of melanin. The effect of all the genes is additive and the amount of melanin produced is always proportional to the number of dominant genes. Davenport found that the two pairs of genes Aa and Bb cause the difference in skin pigmentation between Negro and Caucasian population. A Negro has four dominant genes AABB, and a Caucasian has four recessive genes aabb. The F_1 offspring of mating of aabb with AABB, are all AaBb and have an intermediate skin colour termed mulatto (Figure 5.4). A mating of two such mulattos produces a wide variety of skin colour in the offspring, ranging from skins as dark as the original Negro parent to as white as the original Caucasian parents.

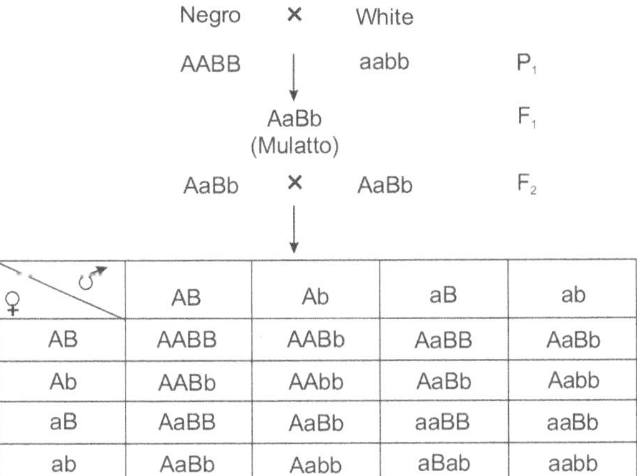

♀ \ ♂	AB	Ab	aB	ab
AB	AABB	AABb	AaBB	AaBb
Ab	AABb	AAbb	AaBb	Aabb
aB	AaBB	AaBb	aaBB	aaBb
ab	AaBb	Aabb	aBab	aabb

Figure 5.4   Inheritance of skin colour in man

F_2 results are shown in the following table.

Phenotypes	Genotypes	Genotypic frequency	Phenotypic ratio
Black (Negro)	AABB	1	1
Dark	AaBB	2	
	AABb	2	4
Intermediate (Mulatto)	AaBb	4	
	aaBB	1	
	AAbb	1	6
Light	Aabb	2	
	aaBb	2	4
White	aabb	1	1

These results clearly indicate that A and B, the dominant genes produce about the same amount of skin pigmentation. Therefore the decrease or increase in number of A and B genes cause different phenotypes. There is gradation in skin phenotypes as black (4 dominant genes), dark (3 dominant genes), mulatto (2 dominant genes), light (1 dominant gene) and white (0 dominant genes). The phenotypic ratio is 1 black : 4 dark : 6 mulatto : 4 light : 1 white.

Height in Man

Skin colour in man is a simple example of polygenic inheritance since only 2 pairs of genes are involved. But the inheritance of height in man is more complex as there are ten or more pairs of genes involved in controlling the trait. In man, tallness is recessive to shortness. So individuals having a genotype of more dominant genes will have shortness as their phenotype. This polygenic trait is controlled by multiple pairs of genes and is variously influenced by a variety of environmental conditions. The heights of adults generally range from 140 cm to 203 cm. If the height of a thousand adult men is measured and the individual height is plotted against height in centimetres and the points connected, a bell-shaped curve is produced and is termed as normal distribution curve and is characteristic of quantitative inheritance.

Summary

- Gene interaction refers to interaction between genes at different loci to produce a single phenotype.
- These interactions produce some totally new trait or phenotype that is not represented in the parents.
- Gene interaction often results in modified dihybrid ratios.
- The epistatic gene present in one locus suppresses the expression of hypostatic gene.
- The complementary genes are two pairs of non-allelic dominant genes which interact to produce only one phenotypic trait, but neither of them when present alone produces the phenotypic trait in the absence of the other.
- Supplementary genes are two independent pairs of dominant genes which interact in such a way that one dominant gene will produce its effect whether the other is present or not.
- Lethal alleles may affect viability as well as the phenotypic traits of an organism.
- Some traits are controlled by polygenes. The phenotypic expression indicates gradation in the character depending on the number of dominant or recessive gene copies present.

Review Questions

1. How does the cross involving fowls with different comb patterns differ from the usual Mendelian cross?

2. Giving reasons, differentiate the Mendelian dihybrid ratio and the ratios for complementary and supplementary interaction.

3. What is the genetic reason behind the production of a creeper fowl?

4. Enumerate the distinguishing features of polygenes.

5. List the gradation of skin genotypes and phenotypes based on the pigmentation produced by the dominant genes A and B.

6

MULTIPLE ALLELES

Objectives

- ✿ To understand the occurrence of more than two alleles in a gene locus controlling a trait
- ✿ To analyse multiple allelic series for skin colour in rodents and eye colour in *Drosophila*
- ✿ To study blood group inheritance and its implications

Key Terms

multiple alleles	agouti	chinchilla
albino	agglutination	agglutinin
agglutinogen	universal donor	codominant
universal recipient	Rh incompatibility	
erythroblastosis foetalis		

INTRODUCTION

The concept of inheritance by Mendel envisaged alternative forms of a same gene. This means that each gene has two alternative forms, the alleles or allelomorphs, one being dominant and the other recessive; one being the wildtype and the other mutant. The mutant gene can give rise to many other mutant genes and therefore a gene can have more than two alleles. These alleles make a series of multiple alleles.

Multiple alleles can be defined as a set of three, four or more alleles or allelomorphic genes, which have arisen as a result of mutation of a normal gene and which occupy the same locus in the homologous chromosomes.

CHARACTERISTICS OF MULTIPLE ALLELES

1. Multiple alleles occupy the same locus within the homologous chromosomes. It means that only one member of the series is present in a given chromosome.

2. Since in a diploid cell only two chromosomes of each type are present, only two genes of the multiple series are found in a cell and also in an individual.

3. Crossing over does not occur in multiple alleles.

4. Multiple alleles control the same character.

5. In the multiple alleles, one is dominant and the rest are recessive or sometimes they are co-dominant.

MULTIPLE ALLELES IN ANIMALS

Some examples of multiple allelic traits in animals are skin colour in rabbits and mice, and eye colour in *Drosophila*. The ABO blood group system in humans is also a good example for multiple alleles.

Skin Colour in Rabbits

In rabbits, four kinds of skin colour are known. Rabbits are accordingly classified as coloured (agouti), chinchilla, Himalayan albino and albino. If homozygous coloured (CC) is crossed with albino ($c^a c^a$), F_1 heterozygous Cc^a individual is coloured showing that coloured is dominant over albino. F_1 heterozygous coloured (Cc^a) gives a $3:1$ ratio in F_2 (Figures 6.1 and 6.2).

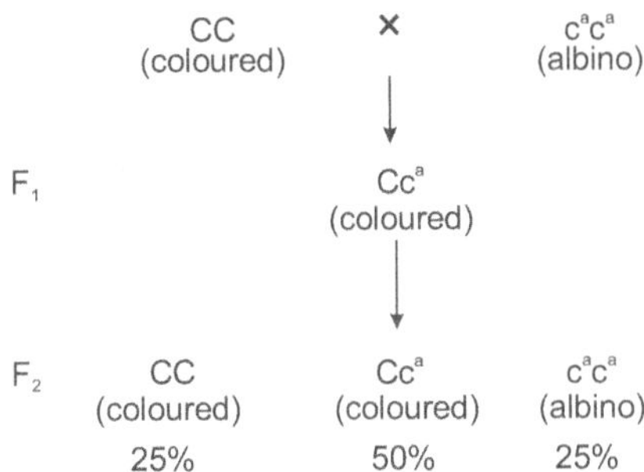

Figure 6.1　Inheritance of skin colour in rabbits from a cross (coloured × albino)

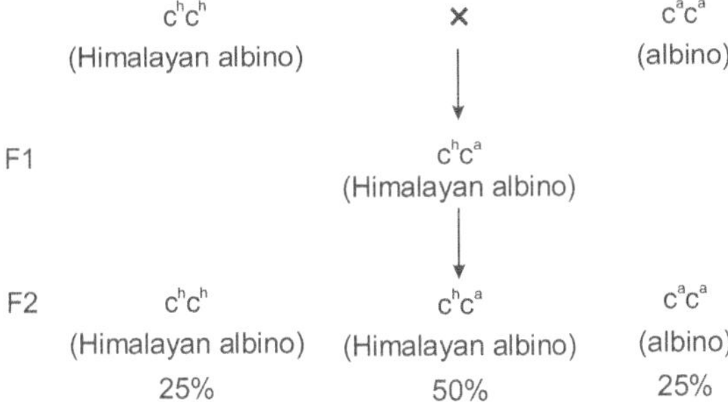

Figure 6.2　Inheritance of skin colour in rabbits from a cross (Himalayan albino × albino)

Still another type is known in rabbits and is called chinchilla. This colour is lighter than the wild agouti. Coloured character is completely dominant over chinchilla. However F_1 hybrids between chinchilla and Himalayan albino ($c^h c^h$) or between chinchilla and albino ($c^a c^a$) show light grey skin colour. A summary of different genotypes involved in the series, along with the other phenotypes is given in Table 6.1.

Table 6.1   Multiple allelic series for skin colour in rabbits

Genotypes	Phenotypes
CC, Cc^h, Cc^a	Coloured (wild)
$c^{ch} c^{ch}$	Chinchilla
$c^{ch} c^h$, $c^{ch} c^a$	Light grey
$c^h c^h$, $c^h c^a$	Himalayan albino
$c^a c^a$	Albino

Skin Colour in Mice

In mice, however, two different multiple allelic series are known for skin colour, one an albino series and the other a black series (Table 6.2). In albino series, it was observed that wild type (grey) was dominant over all others, while heterozygous between any two of the three recessive alleles (a, a^m, and a^e) gave intermediate phenotype (Table 6.3). In black series on the other hand, black is recessive to all, while yellow is dominant to all. Light belly in grey and black (G^L and b^L) is dominant over darker belly. Black body of b^L, however, is recessive to grey body of normal individuals. Consequently, $+/G^L$ and $+/b^L$ both would have grey body and light belly (Table 6.4).

Table 6.2   Multiple allelic series for skin colour in mice

Albino series		Black series	
Genotype	Phenotype	Genotype	Phenotype
+/+	Grey	+/+	Normal grey
a/a	Albino	b/b	Black
a^m/a^m	Medium light	Y/Y	Yellow (lethal)
a^e/a^e	Extreme light	G^L / G^L	Grey with light belly
		b^L /b^L	Black with light belly

Table 6.3   Various combinations for albino series in mice

Genotype	Phenotype
$+/a$, $+/a^m$, $+/a^e$	Grey
a/a^m	Intermediate between a/a and a^m/a^m
a/a^e	Intermediate between a/a and a^e/a^e
a^m/a^e	Intermediate between a^m/a^m and a^e/a^e

Table 6.4   Various combinations for black series in mice

Genotype	Phenotype
$+/b$	Grey
$+/Y$, Y/b	Yellow
$+/G^L$, $+/b^L$, G^L/b^L, G^L/b	Grey body with light belly
b^L/b	Black body with light belly

Eye Colour in *Drosophila*

Normal eye colour in *Drosophila* is red. White-eyed *Drosophila* was one of the first mutations known in *Drosophila*. Red eye and white eye showed simple dominant and recessive relationship.

Subsequently, different shades between red and white were observed. About twelve different allelomorphs are known at this locus. Some of them are shown in Table 6.5. Red eye is dominant over all the other alleles. When any two different alleles were brought together, intermediate type was obtained.

Table 6.5   Multiple allelic series for eye colour in *Drosophila*

Allele	Phenotypes in homozygous condition	Heterozygous condition	
		Genotype	Phenotype
+	Red	+/w	Red eye
w	White	+/w^{ch}	Red eye
w^{ch}	Cherry	+/w^{h}	Red eye
w^{h}	Blood	+/w^{e}	Red eye
w^{e}	Eosin	+/w^{apr}	Red eye
w^{apr}	Apricot	+/w^{iv}	Red eye
w^{iv}	Ivory	+/w^{cr}	Red eye
w^{cr}	Cream	Other heterozygotes	Intermediate

ABO BLOOD GROUPS IN HUMANS

The ABO blood group is an excellent example for multiple alleles. On the basis of presence or absence of certain antigens, ABO blood groups have been established in man by K. Landsteiner. Blood from an individual cannot always be safely mixed with that of another. This is due to the fact that blood proteins of one individual differ from those of others. There are two types of substances in the blood known as antigens or agglutinogens and antibodies or agglutinins. The antigens A and B are located in the RBC and the antibodies a and b in the plasma. If a person has antigen A then the antibody b is present in the plasma. The antigens are found to be mucopolysaccharides. The clumping reaction occurs between antigens and antibodies and is also termed as agglutination (Figure 6.3).

Depending upon the presence of antigens and antibodies, four blood groups have been differentiated. These are designated as A, B, AB and O blood groups. The persons belonging to blood group A have antigen A in their red blood cells and antibody b (anti-B) in their plasma. Persons with blood group B have the antigen B in their red blood corpuscles and antibody a (anti-A) in their plasma. Persons with group AB have antigens A and B in the blood cells and no antibodies in the plasma. Persons with group O have no antigens in their blood cells but both a and b antibodies in their plasma. Table 6.6 represents the presence of antigens and antibodies in different blood groups.

Table 6.6 Distribution of antigens and antibodies in blood groups

Blood groups	Antigens in RBC	Antibodies in plasma
A	A	b (anti-B)
B	B	a (anti-A)
AB	A and B	None
O	None	a and b

Type of blood group	Genotype	Antibodies made by body	Reaction to added antibodies	
			Anti-A	Anti-B
A	I^AI^A or I^Ai	anti-B		
B	I^BI^B or I^Bi	anti-A		
AB	I^AI^B	Neither anti-A nor anti-B		
O	ii	Both anti-A and anti-B		

Figure 6.3   ABO blood group system
(*Source.* biology 200.gsu.edu)

The agglutination reaction has significance in blood transfusion. Persons with blood group O lack antigens and their blood is not clumped by the serum of any other blood group, so that person of O group can give blood to all; but can receive only from O blood group. Hence they are called universal donors. The serum from AB blood group individuals does not cause clumping with any group. Hence they can accept blood from persons with any blood group but can donate blood only to AB blood group persons. So they are known as universal recipients (Table 6.7).

Table 6.7   Blood group compatibility

Donor's group	Can give to	Can receive from	Remarks
O	O, A, B, AB	O	Universal donor
A	A & AB	O, A	-
B	B & AB	O, B	-
AB	AB	O, A, B, AB	Universal recipient

Inheritance of ABO Blood Groups

Bernstein discovered that the inheritance of ABO blood groups is by multiple alleles, I^A, I^B and i. The I^A and I^B genes are co-dominant and so are equally expressed. On the other hand, the i gene is recessive to both the genes I^A and I^B (Table 6.8).

Table 6.8   Genotypes of blood groups

Blood group	Genotypes
A	$I^A I^A$ or $I^A i$
B	$I^B I^B$ or $I^B i$
AB	$I^A I^B$
O	i i

Blood Groups and Disputed Parentage

If the parent blood groups are known, the blood groups of their children can be predicted. Blood groups sometimes help to decide cases of disputed parentage in criminal courts, because a particular pair of blood groups in parents may give some and not all blood groups in progeny. Therefore if a child has a blood group, which is not likely to result from the blood groups of a married couple claiming parentage, then it is proved that the child has a doubtful parentage. Various possible pairs of blood groups of the parents and their children are given in Table 6.9.

Table 6.9   Blood groups of parents and the possible types in children

Parent's blood group	Blood group in offspring	Improbable
O × O	O	A, B, AB
O × A	A, O	B, AB
O × B	B, O	A, AB
O × AB	A, B	O, AB
A × A	A, O	B, AB
A × B	A, B, AB, O	None
A × AB	A, B, AB	O
B × B	B, O	A, AB
B × AB	A, B, AB	O
AB × AB	A, B, AB	O

Subdivisions of the Blood Group A

The blood group A has been divided into subgroups A_1, A_2, A_3 and A_4 depending on the presence of specific antigens and its reaction with B-serum. The subgroup A_1 is common. Each of the subgroups of A is determined by a separate gene and all these genes are alleles. The genes A_2, A_3, and A_4, usually occur in combination with A_1, and A_1 is dominant over all the other A subtypes.

In a person of blood group AB, there might be many of these subtypes present and therefore AB blood group can be divided into subgroups A_1B, A_2B, A_3B, and A_4B.

MN BLOOD GROUPS

Landsteiner and Levine discovered antigens M and N in human blood, which when injected into rabbits stimulated antibody production in them. Human population can be divided into blood groups M, N and MN depending on the presence of antigens M, N and MN respectively, but their serum does not contain antibodies. MN blood groups are controlled by alleles M and N which are equally dominant and so they are co-dominant. Human populations can be divided into three groups on the basis of reaction with anti-M serum and anti-N serum. The alleles for MN blood groups are L^M and L^N (Table 6.10).

Table 6.10   MN blood groups and their genotypes

Blood group	Genotype	Antigen	Antibody	Agglutination	
				Anti-M	Anti-N
M	$L^M L^M$	M	None	+	−
MN	$L^M L^N$	M, N	None	+	−
N	$L^N L^N$	N	None	+	−

RH BLOOD GROUP

IN 1939, Levine and Stetson had implicated an unknown foetal blood group antibody in the haemolytic disease affecting newborn infants. The disease was a common cause of infant mortality during the years 1948–1957. There were 1991 cases reported in Northeast England. Before effective treatment of infants began, almost all of these babies died before they reached the age of one month. At about the same time a new blood group was observed by Landsteiner and Wiener by injecting red cells of rhesus monkey *Macaca rhesus* into rabbits and guinea pigs. The antibodies isolated from them agglutinated with red cells of human. The red cells of the monkeys contain the antigen Rh. In humans it was found out that some persons possess Rh antigen and others do not have it. Those who have the antigens are termed as Rh-positive and those who do not have it as Rh-negative.

According to Fischer there are three dominant genes, C, D and E, and three recessive alleles, c, d and e, determining Rh-positive or negative blood groups. If the alleles combination is cde, then the blood group is Rh-negative. On the other hand, Rh-positive blood group is determined by the following combinations, Cde, CDe, CdE, CDE, cDE, CDe and cdE.

The Rh factor has great significance in child birth. A serious problem occurs, when father is Rh-positive and mother is Rh-negative. In such cases all children born will be Rh-positive, if the father is homozygous. Similarly if the father is heterozygous positive then half of the children are Rh-positive.

Rh incompatibility can cause the disease erythroblastosis foetalis (Figure 6.4). Rh blood group incompatibility occurs when a woman who is Rh-negative marries a man who is Rh-positive. The first child of such a marriage is usually normal. However, during the following pregnancies the foetus may be lost by stillbirth or it may be born in such an anaemic or jaundiced state that it lives only a few hours or days after birth. The infant dies from erythroblastosis fetalis, a condition of anaemia due to the breakdown of RBC (haemolysis) in the foetus, and jaundice. The blood vessels in the liver become clogged with the broken RBCs and the condition is called haemolytic jaundice. This is in response to destruction of baby's RBC by the maternal Rh antibodies. In some cases, babies may be stillborn or may die shortly after delivery.

The blood vascular system of mother and child are quite separate. Any exchange of nutrients or other metabolic products between mother and child takes place by diffusion across the placenta. Placenta is an effective barrier which restricts the passage of most substances other than those directly involved in the normal development and growth of the offspring. However at times, capillaries of the placenta may become defective so that they break or allow the seepage of blood from the foetus into that of the mother. If Rh-positive blood from foetus enters the circulatory system of Rh-negative mother, the response is the formation of antibody against Rh-positive blood. The antibodies from the mother may then pass back across the placenta. Generally, the antibodies produced on the first pregnancy are not sufficient to produce serious effect. The second pregnancy will increase the concentration of antibodies in mother's blood due to immunization. In this case the foetus RBCs will be destroyed causing erythroblastosis foetalis.

A newborn infant with erythroblastosis foetalis may be given blood transfusion to replace the blood with damaged RBCs. Antibodies against Rh antigen namely Rh immunoglobulin (Anti-D) is given to the mother within 36 hours of delivery of child (Figure 6.4). This will destroy any Rh-positive foetal red cells that might have entered maternal circulation and prevent Rh sensitization against later Rh-positive pregnancy (Figure 6.5).

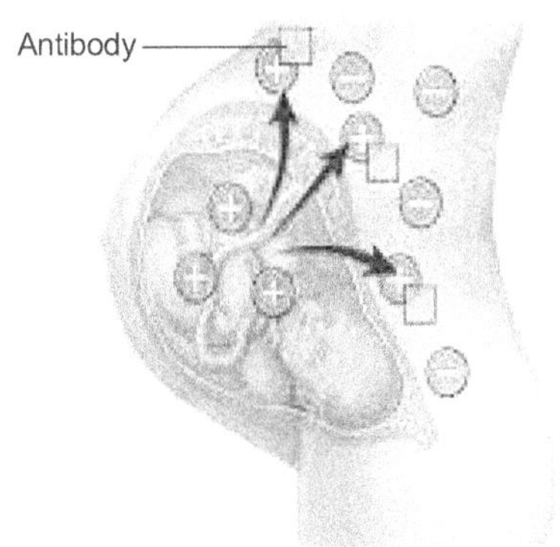

Figure 6.4　Rh incompatibility
(*Source*: www.edward.org)

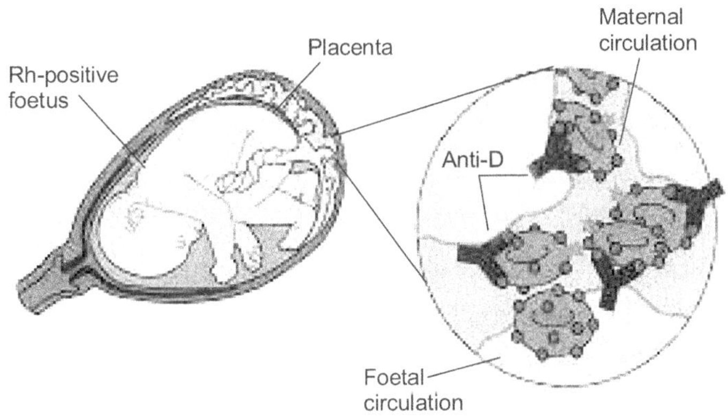

Figure 6.5　Protection against erythroblastosis foetalis
(*Source*: Adapted from Mangae, E.J. and Mangae, A.P., 1998.)

Summary

- Multiple alleles are a group of alleles residing in the same locus, responsible for a trait in contrast to a pair proposed by Mendel.
- The skin colour in mammals and the eye colour in *Drosophila*, are controlled by multiple alleles.
- ABO blood group is another example for multiple alleles.
- Blood group incompatibility is evident by the antigen–antibody reaction which depends on the genotype.
- The inheritance of the blood group and the prediction of parent and offspring blood groups are possible by analysing the genotypes.
- MN blood group system is an example for co-dominant genes.
- Rh blood group is controlled by a series of alleles.
- Rh incompatibility leads to erythroblastosis foetalis.

REVIEW QUESTIONS

1. Enumerate the features of multiple alleles.

2. What are the multiple allelic series responsible for the coat colour in rabbits? State their phenotypes and genotypes.

3. Work out the possible phenotyes and genotypes in the offspring born to parents of A type and B type blood group.

4. Give the genetic reason for the occurrence of erythroblastosis foetalis.

5. What is Rh incompatibility and how is it treated?

6. How is it possible for two Rh-positive parents to have an Rh-negative child?

7. A woman of blood type O, Rh-positive, has a child of blood type A, Rh-negative. What are the possible genotypes of the father for the ABO and Rh blood groups?

7

LINKAGE AND CROSSING OVER

Objectives

- ✿ To understand that the genes located on the same chromosome do not assort independently but are linked together and inherited together
- ✿ To study the recombinant arrangements of genes produced by crossing over

Key Terms

linkage	coupling	repulsion
cis-arrangement	*trans*-arrangement	crossing over
synapsis	bivalent	chiasmata
tetrad	terminalization	chromatid
zygotene	diplotene	carnation eye
bar eye	recombination	

INTRODUCTION

Mendel proposed his laws of segregation and independent assortment by observing progeny of genetic crosses. Mendel never had the idea of the biological process that produced these phenomena. In 1903, **Walter Sutton** proposed a biological basis for Mendel's principles, called the **chromosomal theory of inheritance**. The features of the chromosomal theory are as follows:

1. The chromosomal theory considers the chromosomes as the organs of heredity, and the genes are found on chromosomes.

2. Chromosomes form the link between two generations.

3. The genetic material localized in the chromosome and its contents are constant from one generation to the next.

4. The chromosomes replicate the genetic information contained in their DNA molecule and this information is transcribed at the right time into specific RNA molecules which direct the different types of proteins for specific functions and to form a body like that of parents.

Mendel's laws can be explained in terms of chromosomal theory of heredity. Each individual possesses two alleles for a trait, each of which is located at the same position, or locus, on each of the homologous chromosomes. These chromosomes segregate in meiosis and each gamete receives one of the homolog. In meiosis, each pair of homologous chromosomes assort independently of the other homologous pair.

LINKAGE

The number of chromosomes in most organisms is limited and the number of genes is more than the chromosomes. Each chromosome carries many genes. These genes being located in the same chromosome cannot assort independently as stated by Mendel, rather these genes tend to be inherited together. This phenomenon of inheritance of genes together, retaining the parental combination in the offspring is known as linkage. These genes are located in the same chromosome and inherited together and are known as linked genes and the characters controlled by these are linked characters. All those genes which are located in the single groups in an organism are equal to the number of chromosome pairs. For example, there are 4 linkage groups in *Drosophila*, 23 in man and 20 in mouse. The theory of linkage was proposed by T.H. Morgan in 1911. But the phenomenon of linkage was predicted and described earlier.

Coupling and Repulsion Hypothesis

Bateson and Punnett observed linkage but described it as **coupling**. They found that the results of dihybrid cross in sweet pea, do not agree with the law of independent assortment. The results obtained are shown below.

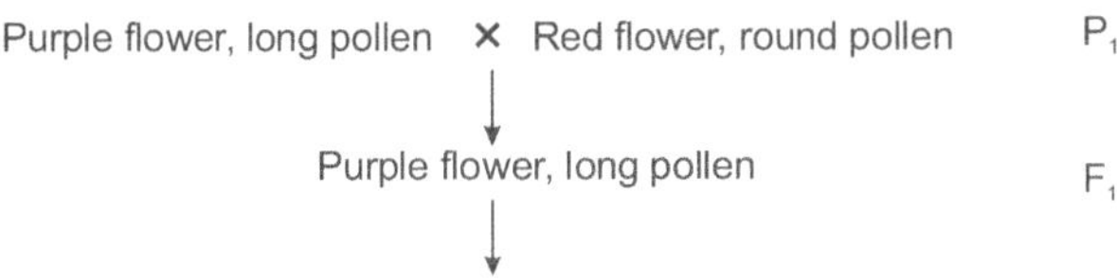

When the F_1 (purple, long, hybrids) were crossed with the double recessive red and round individuals (test cross) they failed to produce the expected 1 : 1 : 1 : 1 ratio in F_2 generation. Instead the following four combinations in the ratio of 7 : 1 : 1 : 7 are produced.

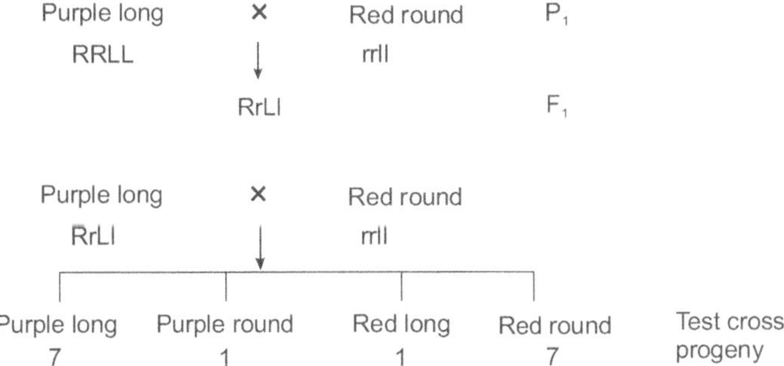

The above results of the test cross indicate that the parental combinations are seven times more numerous than the non-parental combinations. Bateson and Punnett suggested that the alleles coming from the same parent tend to enter the same gamete and tend to be inherited together called genetic coupling. Similarly, the same genes coming from two different parents tend to enter different gametes and tend to be inherited separately and termed as repulsion.

Morgan's Concept of Linkage

T.H. Morgan while working on *Drosophila* stated that coupling and repulsion are two aspects of the same phenomenon, which he described as linkage. He defined linkage as the tendency of the genes, as present in the same chromosome, to remain in their original combination and enter together in the same gamete.

The theory of linkage proposed by T.H. Morgan and Castle has the following features.

1. The linked genes are present in the same chromosome.
2. Linkage of genes is linear.
3. Distance between the linked genes is inversely proportional to the length of the genes.
4. Linked genes remain in their original combination during the course of inheritance.

Arrangement of Linked Genes

In an individual heterozygous for two pairs of linked genes, the linkage can be either of the following two types.

1. ***cis*-arrangement**   The dominant genes of both the pairs are located in the same chromosome and their recessive alleles in the other member of the pair (AB/ab).

2. ***trans*-arrangement**   The dominant gene of one pair and the recessive gene of another pair are located on one chromosome (Ab/aB).

Linkage in *Drosophila*

A cross between wild type *Drosophila* with grey body and vestigial wings (b⁺b⁺vg vg) and *Drosophila* with black body and long wings (bbvg⁺vg⁺) produces F_1 offspring, all of which have grey body and long wings (b⁺bvg⁺vg). These F_1 male hybrids when back-crossed with a double-recessive female (test cross) having black body and vestigial wings (bbvgvg) produce offspring of two types in equal proportion (Figure 7.1). These offspring resemble the two grandparents.

The results indicate that grey body character is inherited together with the vestigial wings. It means that these genes are linked together. Similarly black body character is associated with the long wings. In the above example, the offspring exhibits only the parental combinations of characters, revealing complete linkage.

CROSSING OVER

According to the theory of linkage the linked genes are located in the same chromosome in a linear fashion. These genes located in one chromosome are inherited together generation after generation and only parental combinations appear. But in some cases, though the parental combinations are numerous, non-parental combinations also appear.

This indicates that linked genes do separate and there is an interchange of alleles resulting in the appearance of non-parental combinations. Morgan termed this phenomenon as crossing over.

Example of Crossing Over

A cross between a grey-bodied, vestigial-winged (b⁺b⁺vgvg) and black-bodied, long-winged (bbvg⁺vg⁺) *Drosophila* produces F_1 hybrids, all of them having grey body and long wings (b⁺bvg⁺vg). When female flies of F_1 generation (b⁺bvg⁺vg) were crossed with double recessive males having black body and vestigial wings (b⁺b⁺vg⁺vg⁺), i.e., when a test cross was performed, four types of offspring were produced as follows:

1. Grey vestigial   41.5% ⎫ 83% Parental
2. Black long   41.5% ⎭   combination

3. Black vestigial   8.5% ⎫ 17% Non-parental
4. Grey long   8.5% ⎭   combination

The parental combinations are the result of linkage and are non-crossover while non-parental combinations are due to crossover or recombination (17%) (Figure 7.1).

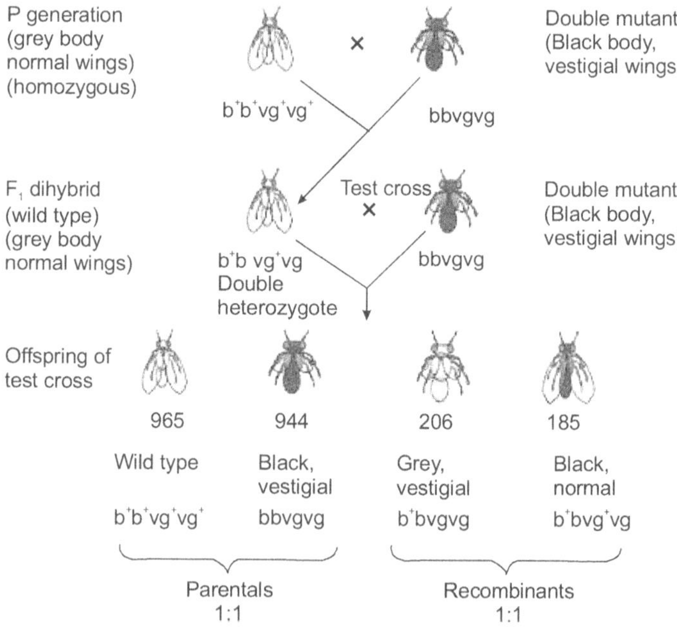

Figure 7.1   Linkage and crossing over in *Drosophila*
(*Source*: www.biolgy.uiowa.edu)

Mechanism of Crossing Over

The behaviour of chromosomes during meiosis describes the mechanism of crossing over.

Synapsis   During prophase I of meiosis, the maternal and paternal chromosomes of a homologous pair come closer together and pair at zygotene stage. This pairing is termed synapsis. The homologous chromosomes pair precisely by mutual attraction between the allelic genes. The paired chromosomes are called bivalents.

Duplication   During diplotene stage of meiosis, each of the homologous chromosomes in a bivalent splits longitudinally into two sister chromatids. Thus the bivalent now consists of four chromatids and is known as tetrad.

Chiasma formation   During diplotene stage, when the paired chromosomes start separating, the chromatids remain in contact at one or more points. These points of contact are known as chiasmata. At each chiasma two non-sister chromatids of the bivalent break at the corresponding points and then rejoin with the exchange of segments (Figure 7.2).

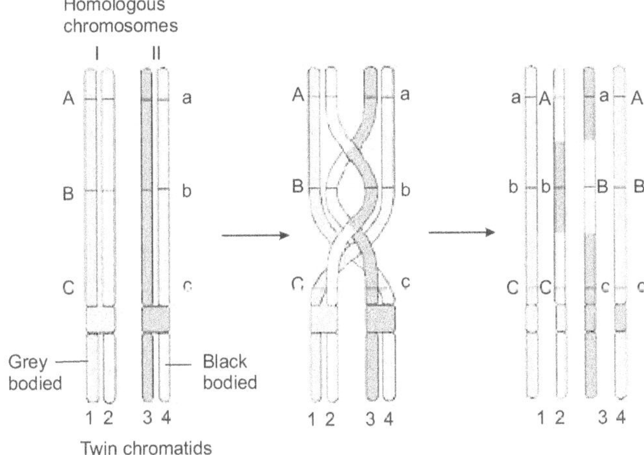

Figure 7.2   Chiasma formation and crossing over
(*Source*: www.biology.online.org)

Terminalization   After crossing over, the non-sister chromatids start repelling each other. The chromatids separate from the centromere towards the tip. The chiasmata also start moving in a zipper-like fashion towards the ends. The movement of chiasmata is known as terminalization.

Kinds of Crossing Over

Depending upon the number of chiasmata, crossing over can be of the following types :

Single crossing over   When only one chiasma is formed all along the length of a chromosome pair, it is known as single crossing over. The chromatids of homologous chromosomes contact and break only at one point along their entire length.

Double crossing over   In double crossing over, chromatids break and rejoin at two points. It means two chiasmata are formed along the entire length of the chromosome. In double crossing over, the formation of each chiasma is independent of the other. It means each crossover can involve either of the two chromatids of each homologous chromosome so that three different types of double crossover tetrads are formed.

i.   **Two-strand double crossover**   In these the same two chromatids are involved at both the crossing over points. The chiasmata thus formed are known as **reciprocal** chiasmata. As a result of this, two chromatids are non-crossovers and preserve the parental combination of genes.

ii. **Three-strand double crossover**   In this cross over, three chromatids are involved, i.e., the second chiamas contains one of the same chromatid which has crossed at first chiasma but the other chromatid is different. Here only one chromatid is non-crossover, while the other three exchange their parts.

iii. Four-strand double crossover   In such crossover, all the four chromatids of the tetrad are involved, two of them exchange parts at first chiasma and the other two are involved in the second chiasma. Such chiasmata are known as complementary chiasmata. They produce four single crossovers.

Multiple crossing over   When crossing over occurs at more than two places in the same chromosome pair and more than two chiasmata are formed, this type of crossing over is known as multiple crossing over.

Features of Crossing Over

1. Only two of the four chromatids crossover, while the other two chromatids preserve their original constitution.

2. The crossing over occurs between non-sister chromatids of the homologous pair of chromosomes.

3. The number of chiasmata per set depends on the length of chromosomes. The longer the chromosomes, the greater is the number of chiasmata.

4. Chances of crossing over are more if the genes are located farther apart. The more the distance between the genes, greater is the opportunity for chiasma formation.

Theories of Crossing Over

Many theories are proposed to explain the mechanism of crossing over.

Darlington's strain theory   C.D. Darlington explained the reasons for breakage of chromatids. According to him, the chromosomes of homologous pair (bivalent) in zygotene stage are rationally coiled around each other. Also the two sister chromatids in each chromosome are coiled one around the other but in a direction opposite the coiling between the homologous chromosomes. When chromosomes start separating, their relational coiling unravels in one direction and the sister chromatids unravel in the opposite direction. The repulsion causes tension which exerts strain on the weak chromatids. The chromatids break at the points of contact and broken ends rejoin forming X-shaped chiasma.

Breakage and exchange theory   This theory was forwarded by Muller and is the most widely accepted theory on the mechanism of crossing over. According to this theory the chromatids undergoing crossing over first break into two or more segments. The broken segments of non-sister chromatids then rejoin with an exchange, forming the chiasma. The enzyme endonuclease causes a break in the non-sister chromatids of homologous chromosomes at the corresponding points and the enzyme ligase governs the reunion of broken non-sister chromatids.

Copy choice theory   Belling has suggested that crossing over is related with the duplication of chromosomes. According to this theory, crossing over occurs during synthesis of new chromatids. The parent chromatids act as templates upon which new chromatids are synthesized. Since the parent chromatids lie close together, the new chromatid is first formed along the paternal chromatid and then switches to the maternal one and thereafter copies the latter. The complementary replica of the maternal strand also switches on to the paternal strand when it reaches the same point. The new chromatids thereby carry information from both chromatids, giving the appearance of having exchanged segments of homologous chromatids.

Cytological Evidence of Crossing Over

In 1931, Curt Stern demonstrated the mechanism of crossing over under the microscope. In this experiment Stern obtained a female *Drosophila* in which the two members of the

X chromosome pair were morphologically different from each other and also from other sets of chromosomes. In this strain one X chromosome had a portion of Y chromosome attached to it, so that it has 'L' shaped appearance (Figure 7.3). This is used as a cytological marker for demonstrating crossing over.

A female with an abnormal X chromosome broken into two segments was made heterozygous for two sex-linked mutations. These genes were made to be present in one broken segment.

1. Carnation eye colour 'c' (a recessive mutant to the normal red 'C' eye colour).

2. Bar shaped eye 'B' (narrowing of eyes dominant to normal round 'b' eyes).

A heterozygous female having genes for carnation coloured and bar shaped eyes (cB) in one X and the wild type red and round eye (Cb) genes in 'L' shaped X was bred to male having both the recessive genes (cb). The offspring of such a cross (test cross) were found to belong to following four groups.

i. Carnation and bar	cB/cb with broken X without any fragment of Y
ii. Red and round	Cb/cb with unbroken X with the attached Y fragment
iii. Red and bar	CB/cb with broken X with the attached Y fragment
iv. Carnation and round	cb/cb with unbroken X without an attached Y fragment

Thus, flies in which crossing over was indicated phenotypically showed microscopic evidence of exchanges between homologous chromosomes. The physical or cytological basis of crossing over was thus established.

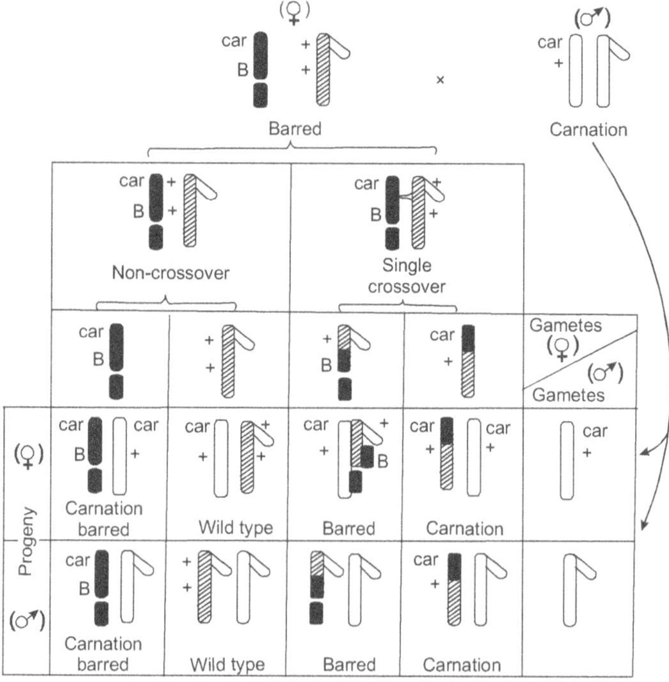

Figure 7.3   Stern's experiment (Adapted from Owen and Edgar, 1965)

Significance of Crossing Over

Crossing over is a universal phenomenon, and is of great significance because of the following:

i. Crossing over has led to the construction of linkage maps or genetic maps of chromosomes.

ii. As a result of crossing over, new gene combinations are produced and they play an important role in the process of evolution.

Summary

- ✿ Genes present in the same chromosome do not assort independently but are linked together and inherited together.

- ✿ Alleles coming from the same parent tend to enter the same gamete and be inherited together. This is called genetic coupling.

- ✿ Similarly, the same genes coming from two different parents tend to enter different gametes and this is termed as repulsion.

- ✿ Crossing over breaks up association between genes and produces recombination.

- ✿ The behaviour of chromosomes during meiosis describes the mechanism of crossing over.

- ✿ Theories have been proposed to explain crossing over. Stern's experiment demonstrated crossing over under the microscope.

- ✿ Crossing over has great significance in the construction of chromosome maps and in the production of new gene combinations, aiding evolution.

REVIEW QUESTIONS

1. How was Mendel's law of independent assortment disproved?

2. Outline the experiment to prove linkage using a cross between grey-bodied, vestigial-winged and black-bodied, long-winged fruit flies.

3. Narrate the stepwise occurrence of crossing over.

4. What effect does crossing over have on linkage?

5. How did Stern provide evidence for crossing over?

8

CHROMOSOME MAPPING IN EUKARYOTES

Objectives

- ✿ To analyse the techniques for locating genes on chromosomes
- ✿ To understand the determination of gene order in a chromosome
- ✿ To learn the unique methodologies applied in human gene mapping

Key Terms

genetic map	centimorgan	lod score
two-point test cross	VNTR	FISH
marker map	flanking marker	thymidine kinase
somatic cell fusion	fluorochrome	microsatellites
deletion mapping	DNA sequencer	interference
duplication mapping	RFLP	three-point test cross
Huntington's disease	chromosome map	
coefficient of coincidence		

INTRODUCTION

The discovery of the phenomenon of linkage and crossing over has established the following facts.

1. The genes are arranged in a linear order on the chromosome.

2. The linkage groups in an organism are equivalent to the number of chromosome pairs, as for example, 4 in *Drosophila melanogaster*.

3. Frequency of crossing over or recombination between genes depends upon the distance between linked genes in a chromosome. The chances of crossing over are more between two distantly placed genes.

4. The strength of linkage is inversely proportional to the distance between linked genes.

5. The position of genes is very definitely in the chromatids.

Based upon the above observations, Strutevant (1913) developed the idea that percentage or frequency of crossing over can be used as a tool to determine the relative distance between the genes in a linkage group and also the order of their arrangement. Strutevant and Morgan plotted the position of the five genes on the X chromosome of *Drosophila*. This graphic representation of genes is now known as chromosome map.

The chromosome map is a condensed, graphical representation of the relative distances between the genes in a linkage group expressed in the percentage of recombination. Chromosome maps calculated by using recombination frequencies are called genetic maps.

CONSTRUCTION OF GENETIC MAP

The method of construction of genetic maps of different chromosomes is called genetic mapping. The steps involved in chromosome mapping are as follows:

Determination of Map Distance

It is necessary to know how many genes remain always together or linked during the course of inheritance. Thus the different linkage groups of species can be worked out.

Once the number of genes in each linkage group is established, the relative distance between each linked gene has to be determined. The distance between two genes is calculated according to the percentage of crossing over because crossover frequency is directly proportional to the distance between the genes. For example, if the percentage of crossing over between two linked genes is 1 per cent, it means that the map distance between two linked genes is one unit of map distance, which is known as map unit, morgan or centimorgan.

Examples

If an F_1 hybrid having the genotype Ab/aB produces 8% of crossover gametes AB and ab, then the distance between A and B is considered to be 16 map units or centimorgan.

If the map distance between the gene loci B and C is 12 centimorgan, then 12% of gametes of genotype BC/bc should be crossover types, i.e., 6% Bc and 6% bC.

This is because, each chiasmata produces 50% crossover products and is equivalent to 50 map units or centimorgans. If the mean number of chiasmata is known for a chromosome pair, the total length of the map for that linkage group may be estimated.

$$\text{Total length} = \text{Mean number of chiasmata} \times 50$$

Two-point test cross   The percentage of crossing over between two linked genes is calculated by test crosses in which an F_1 dihybrid is crossed with a double-recessive parent. Such crosses involve crossing over at two points and so called two-point test crosses. For example, a dihybrid having the genotype AC/ac is test-crossed with a double-recessive parent ac/ac, then among the F_2 test cross hybrids, the parental type gametes produced were 37% (AC/ac) and 37% (ac/ac). The crossover gametes (non parental) were 13%, Ac/ac and 13% ac/ac. Thus 26% of all gametes were of crossover types and the distance between the loci A and C is estimated to be

26 centimorgans. For genes located farther apart, three-point test crosses are used.

Three-point test cross   A three-point test cross or trihybrid test cross (involving three genes) gives information regarding relative distance between these genes, and also shows the linear order in which these genes are present in the chromosome. If, in addition to genes A and C, a third marker gene B is located in the same linkage group, all the three can be used together for map distance analysis and to find the relative position of three points.

A test cross is performed between the trihybrid individuals of genotypes ABC/acb. The following progeny are obtained.

36% ABC/abc	9% Abc/abc	4% ABC/abc	1% Abc/abc
36% abc/abc	9% aBc/abc	4% abc/abc	1% aBc/abc
72% Parental type	18% Single crossover between A and B (Region I)	8% Single crossover between B and C (Region II)	2% Double crossover

To find the distance A–B crossover, both single and double that occurred in region I are counted. That is 18% + 2% = 20% or 20 map units between the loci A and B. To find the distance B–C, all crossovers, both single and double that occurred in region II are considered. That is, 8% + 2% = 10% or 10 map units between the loci B and C. The A–C distance is therefore 30 map units.

Determination of Gene Order

After determining the relative distances between the genes of a linkage group, it is easier to place genes in their linear order. For example, if the linear order of three genes A, B and C is to be determined, then those three genes may be in any one of three different orders depending upon which gene is in the middle. If we suppose that the distance A–B = 12, B–C = 7 and A–C = 5, we can determine the order of genes correctly in the following manner.

Case 1   It is assumed that gene A is in the middle, for example, B–A–C.

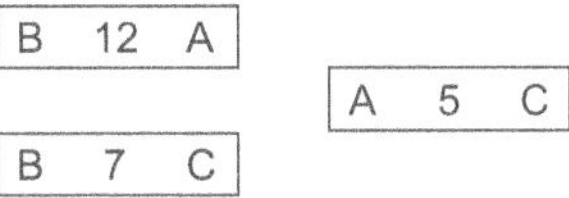

In this case, because the distances between B and C are not equitable, gene A cannot be in the middle.

Case 2   It is assumed that gene B is in the middle. For example, A–B–C.

In this case, because the distances between A and C are not equitable, therefore, gene B cannot be in the middle.

Case 3   It is assumed that gene C is in the middle. For example, A–C–B.

In this case, because the distances between A and B are equitable, therefore, gene C must be in the middle.

Thus, the relative distances and ordering of genes in a linkage group are determined in separate segments by two-point test crosses or three-point test crosses.

Combining Map Segments

Finally, the different segments of maps of a complete chromosome are combined to form a complete genetic map, 100 centimorgans long for a chromosome.

The three map segments can be combined as follows:

I	a 8 b 10 c

II	c 10 b 22 d

III	c 30 e2d

Each of these segments can be superimposed by aligning the genes shared in common.

I	a 8 b 10 c

II	d 22 b 10 c

III	d2e 30 c

Then finally the three segments can be combined into one map.

The distance a to d = (d to b) − (a to b) = 22 − 8 = 14

The distance a to e = (a to d) − (d to e) = 14 − 2 = 12

d2e 12 a 8 b 10 c

INTERFERENCE AND COINCIDENCE

In most higher organisms it has been found that one chiasma formation reduces the probability of another chiasma formation in an immediately adjacent region of the chromosome, probably because of physical inability of the chromatids to bend back upon themselves within certain minimum distances. The tendency of one crossover to interfere with the other crossover is called interference.

The net result of interference varies in different segments of the chromosome and is usually expressed in terms of coefficient of coincidence, i.e., the ratio between the observed and the expected double crossover.

$$\text{Coefficient of coincidence} = \frac{\%\,\text{Observed double crossovers}}{\%\,\text{Expected double crossovers}}$$

When interference is complete that is 1.0, no double crossovers will be observed and coincidence becomes zero. When interference decreases, coincidence increases. Coincidence values ordinarily vary between 0 and 1. Coincidence is generally quite small for short map distance.

CYTOLOGICAL OR PHYSICAL MAPPING

In cytological mapping, localization of genes has been accomplished by identifying gene locus in relation to some visible landmark as chromosome bands. Bridges in 1934 used polytene chromosome of *Drosophila* for cytological mapping. about 5000 single crossbands have been noted on the four pairs of salivary gland chromosomes in *D. melanogaster*. This number is considered approximately as the number of genes in that animal. Bridges system of designating parts of chromosomes with subdivisions had made it possible for investigators to discuss precise locations. In this system, uniform divisions are numbered in order throughout the entire chromosome set from 0 at the beginning of the X chromosome to 102 at the end of chromosome 4. Subdivisions within the areas are identified with letters from A to F and bands within subdivisions are numbered from left to right. For example, the gene (w) for white eyes is in bands 3C2. In linkage map the gene is located at 1.7 in the X chromosome.

Differences Between Genetic and Cytological Maps

The frequency of crossing over usually varies in different segments of the chromosome, but is a highly predictable event between any two gene loci. The actual physical distances between linked genes bears no direct relationship to the map distances calculated on the basis of crossover percentages. However, the linear order or linked genes is identical in both maps.

GENE MAPPING IN HUMANS

Mapping human genome is difficult because of the inability to perform desired crosses and the small number of progenies. Information gathered from analysing pedigrees is often used to determine linkage. To determine the probability that genes are actually linked, geneticists often calculate lod (logarithm of odds) scores. To obtain lod score, the probability of obtaining the observations both with a specified degree of linkage and with independent assortment are calculated. Then the ratio of these probabilities is determined and the logarithm of this ratio is the lod score. Suppose that the probability of obtaining a particular set of observations with linkage and a certain recombination frequency is 0.1 and the probability of obtaining the same observations with independent assortment is 0.0001, the ratio of these probabilities is 0.1/0.0001 = 1000. The logarithm of this is 3. Thus linkage with the specified recombination is 1000 times as likely to produce what was observed as independent assortment. A lod score of 3 or more is usually considered convincing evidence for linkage.

DNA Marker Maps

DNA markers have proved very useful in linkage studies, and about 10,000 such sites have been positioned on the human chromosomes. The first DNA markers that were extensively used are called RFLPs (Restriction fragment length polymorphisms). Others are microsatellites, very short units repeated 20 to 30 times, and VNTRs (variable nucleotide tandem repeats), somewhat longer repeats that are the basis of some DNA fingerprinting.

Restriction fragment length polymorphism is based on substitution mutations that alter the recognition sites of restriction enzymes. Thus, at those particular spots, the enzyme cuts one person's DNA but not another's. For example, suppose that in one person's DNA, there are four for the restriction enzyme *Eco*RI, restriction sites are spaced as follows:

```
--X----------X------X----------X----
      7 kb       1 kb      4 kb
```

The same section in another person's DNA has only three *Eco*RI sites.

```
--X--------------X---------X----
        8 kb          4 kb
```

The difference is due to mutations and most of them are not in protein-encoding genes. The variations they produce can be detected by DNA fingerprinting.

Huntington's disease was the first disease producing gene mapped by DNA markers. The gene was located to an RFLP site about 4 map units away near the tip of the short arm of chromosome 4. Later researchers found DNA markers within 1 or 2 map units on both sides of the gene. These are called flanking markers. Similarly the breast cancer gene *BRCA*1 is within 10 map units of a VNTR with multiple alleles in band q21 of chromosome 17.

The frequency of crossover is near the ends of the chromosomes. On an average, the approximately 3×10^9 base pairs of a haploid genome are divided into 3,700 units for all human chromosomes. Thus $(3 \times 10^9)/3,700$, or approximately 800,000 base pairs, correspond to 1 map unit. So if a disease gene is flanked by known DNA markers, each 1.25 map units away, the gene resides in a DNA interval of 2 million base pairs.

Physical Maps

Physical maps are based on cytological landmarks on the chromosomes like rearrangements, duplications, etc. Physical mapping deals with sequencing lengths of DNA by ordering individual bases A, T, G and C.

The technique called somatic cell fusion is applied in cytological mapping. The laboratory-cultured human cells and mouse cells are fused to form hybrid cells by the fusion agents like PEG or Sendai virus. In such hybrid cells it is interesting to note that all the mouse chromosomes remain while some human chromosomes are lost. The reason for preferential loss is not clear. Which particular human chromosomes are lost and which ones are retained seems to be random. Based on this, sub-clones could be produced.

The loss of the chromosome is often accompanied by loss of a protein encoded by the gene. By associating the gene product with the chromosome, the location of the gene can be predicted. For example, the enzyme thymidine kinase is absent in the cells where the chromosome 17 is lost or whenever the cell contains chromosome 17, the thymidine kinase enzyme is always present.

Another method for gene localization is FISH (Fluorescent *in situ* hybridization). In this technique, first human metaphase chromosomes are prepared. After removing RNA and proteins, single-stranded DNA is produced. These DNA are bathed with a single-stranded probe specific for a given gene. This probe is chemically linked to fluorochrome, a molecule that illuminates under UV rays. Using different fluorochromes, multiple-coloured chromosomes, chromosome painting could be displayed. Thus the chromosomes are painted and could be labelled with a specific gene sequence. By FISH technique, spatial organization of the genes could be observed.

Mapping by DNA Sequencing

The ultimate physical map of human DNA is the specification of base-by-base determination using the common Sanger method. The human DNA (10,000 to 40,000 bases) is inserted in a plasmid, phage or artificial chromosome by breaking up the DNA into 300 to 500 base pieces and order them by the overlapping the regions. Advances in computer-based methods for collecting, storing, distributing and analysing are also done in automated DNA sequencer. With these methods, complete physical maps of entire human genomes have been produced.

Summary

- Construction of gene map is based on the percentage of recombination between two linked genes. Test crosses are conducted to determine recombination frequency.

- The different segments of maps of a complete chromosome are combined to form a complete genetic map.

- The tendency of one crossover to interfere with the other crossover is called interference.

- The net result of interference varies in different segments of the chromosome and is usually expressed in terms of coefficient of coincidence.

- Gene mapping in humans can be accomplished by a variety of techniques.

- DNA markers have proved very useful in linkage studies.

- Physical maps are based on cytological landmarks on the chromosomes like rearrangements and duplications.

- The technique called somatic cell fusion is applied in cytological mapping. By FISH technique spatial organization of the genes could be observed.

- The ultimate physical map of human DNA is the base-by-base specification by DNA sequencing.

REVIEW QUESTIONS

1. How is map distance determined?

2. Mention the purpose of three-point test cross.

3. State the relationship between interference and coincidence.

4. In fruit flies three genes are linked in one chromosome. Assume one parent is dominant for all three genes, the other recessive. In a test cross, the following numbers were obtained. ABC-225, abc-245, aBc-98, AbC-102, ABc-144, abc-156, aBc-14, Abc-16 and total = 1000. What is the crossover percentage? Arrange the genes in linear order.

5. How will you assign human genes in a chromosome by human–mouse cell hybrids?

6. Name the gene markers employed in human gene mapping.

9

SEX DETERMINATION AND SEX-LINKED INHERITANCE

Objectives

- ✪ To analyse the different kinds of sex-determining mechanisms in animals and the factors associated with it
- ✪ To know about sex determination in humans
- ✪ To understand the phenomenon of dosage compensation with respect to sex chromosomes
- ✪ To learn the inheritance pattern of traits that are linked to the sex chromosomes
- ✪ To study the occurrence of traits limited to one sex and traits influenced by one sex

Key Terms

hermaphroditism	monoecious	dioecious
allosomes	autosomes	syndrome
lygaeus	homogametic	heterogametic
protenor	haplodiploidy	freemartin
sex chromatin	sex switch	sex-lethal
non-disjunction	Barr body	gynandromorphs
blastomere	sex reversal	andrase
gynase	histocompatability	Y antigen
genic balance theory	dosage compensation	
androgen-insensitivity	testis–determining factor	
ovary-determining factor	sex-determining region Y (SRY)	

INTRODUCTION

Sexual reproduction is the formation of offspring that are genetically distinct from their parents and both the parents contribute genes to their offspring. Sexual reproduction consists of two processes and there is an alternation of haploid and diploid cells: meiosis produces haploid gametes, and fertilization produces diploid gametes.

The term sex refers to sexual phenotype. Most eukaryotes have two sexual phenotypes: male and female. Biologically, sex is an aggregate of morphological, physiological and behavioural qualities that differentiate the organisms producing eggs from those organisms producing sperm. Egg-producing organisms are known as females and those producing sperms are males. There are many ways in which sex differences arise. In some species, both sexes are present in the same individual, a condition termed hermaphroditism; organisms that bear both male and female reproductive structures are said to be monoecious. Species in which an individual has either male or female reproductive structures are said to be dioecious. The sex behaves as a Mendelian character. Its inheritance follows the law of segregation. The inheritance of traits with regard to sex is widely studied and has applications in human genetics.

The mechanism by which sex is established is termed as sex determination. There are various mechanisms operating in determining the sex of different organisms. The following are the major kinds of sex-determination mechanisms.

1. Chromosomal sex-determining systems
2. Genic balance system
3. Hormonal system
4. Environmental sex determination

MECHANISMS OF SEX DETERMINATION

CHROMOSOMAL THEORY OF SEX DETERMINATION

The chromosomal theory of inheritance states that genes are located on chromosomes, which serve as the vehicles for gene segregation in meiosis. In majority of animals a pair of sex chromosomes are found. These are represented by X and Y.

In 1891, Henking first observed X chromosome in bug. The chromosomal theory of sex determination was given by E.B. Wilson and Stevens (1902). In an organism, X and Y chromosomes are called sex chromosomes or allosomes and the rest of the non-sex chromosomes are somatic chromosomes or autosomes.

Types of Chromosomal Mechanisms of Sex Determination

XX–XY type or Lygaeus type   In many species, the cells of the males and females have the same number of chromosomes but the females have two X chromosomes and the males have one X and a small Y chromosome. XX–XY type of sex-determining mechanism was first studied in the bug *Lygaeus* by Wilson and Stevens. There are two different patterns of Lygaeus type.

Female homogametic and male heterogametic   The females are homogametic sex with XX chromosomes. They produce eggs of same type, carrying only X chromosome. On the other hand, males are heterogametic with XY chromosomes. Males produce two types of sperms, 50% of which possess X chromosome and the other 50% Y chromosome.

Example   All mammals including man, reptiles, *Drosophila* and certain insects.

Drosophila   In *Drosophila* total number of chromosomes are eight, of which six are autosomes, common to both male and female. The other two are sex chromosomes. In male this is represented by XY. Females produce ova. They are all similar possessing 3A+ X chromosomes, whereas the sperm produced by male are 3A+X and 3A+Y in equal numbers [A = Autosomes] (Figure 9.1).

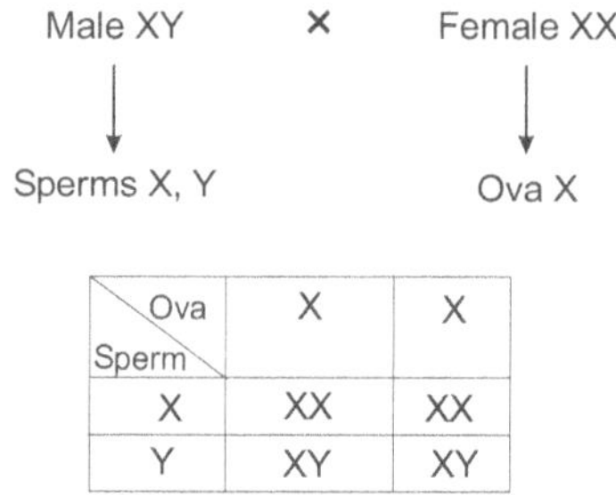

Figure 9.1   Sex determination in *Drosophila*

In case of man total number of chromosomes is 23 pairs or 46.

Male   44A + XY

Female   44A + XX

The sperm produced by male are of two types: 22A + X and 22A + Y, whereas females produce ova all with 22A + X chromosomes (Figure 9.2).

Female heterogametic and male homogametic   In fowls, other birds and some fishes, some amphibians, moths and butterflies, the female sex is heterogametic with X and Y chromosomes. To prevent confusion with XX–XY system, the sex chromosomes in this system are labelled as Z and W but the chromosomes do not resemble Zs and Ws. Females in this system are ZW and laying two types of eggs, 50% with Z(X) chromosome and 50% with W(Y) chromosome. The male sex is homogametic having ZZ(XX) chromosomes. It produces sperms, all of the same type.

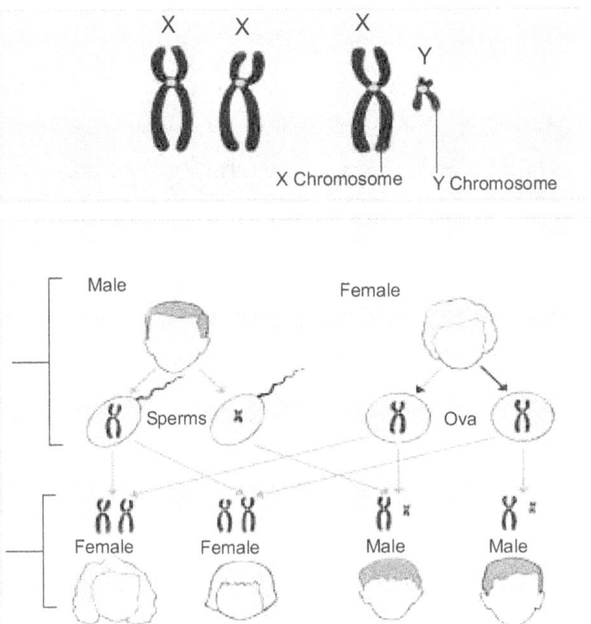

Figure 9.2   Sex determination in man
(*Source*. www.nature.com)

XX–XO type or protenor type   In squash bug it was observed that males have 21 chromosomes, but females have 22 chromosomes. Thus all the eggs are of same type carrying X chromosome while the sperms are of two types; 50% with eleven chromosomes and 50% with ten chromosomes. Fertilization of an egg by a sperm carrying eleven chromosomes results in a female, while egg fertilized by a sperm with ten chromosomes produces male. This type is seen in Orthoptera. Example is grasshopper (Figure 9.3).

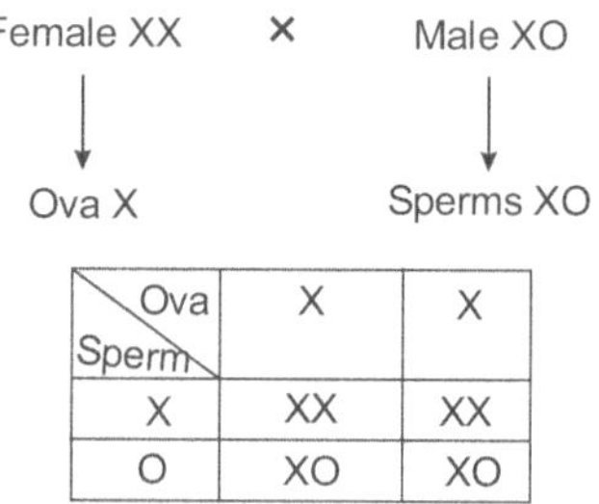

Figure 9.3   Sex determination in grasshopper

HAPLODIPLOIDY MECHANISM

In honeybees, wasps, ants and certain other hymenopterans, parthenogenesis (without fertilization) is widespread. Three kinds of individuals are produced.

1. *Diploid queens*   These develop from fertilized eggs with 32 chromosomes and are fully developed functional females.

2. *Diploid workers*   These also develop from fertilized eggs (32 chromosomes) but are non-functional females and are sterile.

3. *Haploid drones or males*   These develop by parthenogenesis from the haploid unfertilized eggs and are functional males. They have 16 chromosomes.

In some species, a bisexual generation from unfertilized eggs alternates with a female generation from fertilized eggs. There are certain forms where males are altogether absent and females are produced by parthenogenesis. These females are always diploid (Figure 9.4).

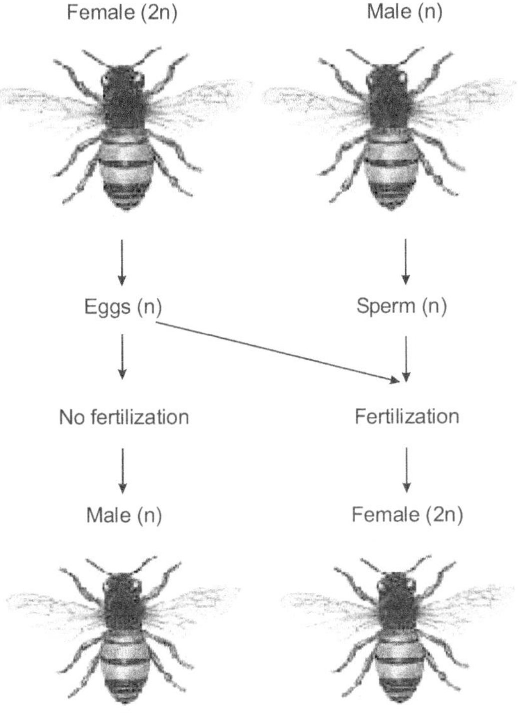

Figure 9.4   Haplodiploidy in honeybee

GENIC BALANCE SYSTEM

Patterson discovered that in *Drosophila*, Y chromosome plays no significant role in sex determination; the ratio between X and the autosomes govern the development of male or female sex. Single dose of X chromosome in a diploid organism produces male whereas 2X chromosomes produce a female.

Bridges worked out the ratio theory of sex determination in *Drosophila*. According to this theory the ratio of X chromosomes to autosomes is the determining factor for sex. If a complete haploid set of autosomes is designated by A then 2A : X will give rise to male and 2A : 2X to female.

Based upon the observations of the ratio theory, Bridges put forward the genic balance theory. According to this theory, every individual whether male or female, possesses in its genotype genes for both male and female characteristics. Depending upon the number of genes of a particular sex the organism's sex will be determined. If there is excess of female-determining genes, a female develops; if the male-determining genes are in excess, a male is formed.

In *Drosophila,* the X chromosome carries more genes that shift the development towards femaleness and the autosomes possess genes which incline the development towards maleness. Therefore, the deciding factor is the ratio between the number of X chromosomes and autosomes.

$$\text{Sex index} = \frac{\text{Number of sex chromosomes}}{\text{Number of sets of autosomes}}$$

For a normal female, sex index is 1. For a normal male, sex index is 0.5. A sex index in between 0.5 and 1 is an intersex. If the sex index is above 1, it is a super female and below 0.5, it is a super male.

Table 9.1 Ratio of X and A chromosomes responsible for determination of sex in *Drosophila*

Sex	Number of X chromosomes	Sets of autosomes	Sex index (X–A ratio)
Super female	XXX	AA	1.5
Normal female	XXXX	AAAA	1.0
	XXX	AAA	1.0
	XX	AA	1.0
Intersex	XX	AAA	0.66
	XXX	AAAA	0.75
Normal male	X	AA	0.5
Super female	X	AAA	0.33

As shown in Table 9.1 the sex index indicates the sex of *Drosophila*.

A sex-switch gene has been discovered that directs female development. This gene, sex-lethal (*Sxl*), is located on the X chromosome and has two states of activity—when it is 'on', it directs female development; when it is 'off', maleness develops. Also, other genes are located on the X chromosome and the autosomes regulate this sex-switch gene. Genes on the X chromosomes that regulate *Sxl* into the 'on' state (female development) are called **numerator elements** because they act on the numerator of the X/A genic balance equation. Genes on the autosomes that act to regulate *Sxl* into the 'off' state (male development) are called **denominator elements**. *Sxl* "counts" the number of X chromosomes; it turns on when two are present. It counts by measuring the level of the numerator gene's protein product. If the level is high, *Sxl* turns on, and the organism develops into a female. If the level is relatively low, *Sxl* does not turn on, and development proceeds as a male.

Intersexes in Drosophila and ratio theory of sex determination   Bridges' hypothesis was supported by studies of flies with abnormal number of chromosomes due to non-disjunction. Due to abnormal meiosis during oogenesis, both the X chromosomes fail to separate, and move to the same pole instead of the opposite poles. This is called **non-disjunction**. So two kinds of abnormal eggs are formed one with AXX and the other with only A without sex chromosome (Figure 9.5). When such abnormal eggs are fertilized with normal sperm, the following results are obtained.

Sperms / Ova	AX	AY
AXX	2AXXX	2AXXY
AO	2AXO	2AYO

AAXXX—Super female
AAXXY—Exceptional female
AAXO—Male
AAYO—Dies

Figure 9.5   Ratio theory in *Drosophila*

Non-disjunction of complete set of chromosomes   The ratio theory of sex determination is further supported by another instance of abnormal meiosis where along with X chromosome, autosomes also fail to separate and move to the same pole of the spindle. Therefore, these abnormal eggs possess AAXX. When such diploid eggs are fertilized with normal sperms the following combinations are produced.

AAAXXX—Triploid female

AAAXXY—Intersex

The intersexes are sterile and intermediate between females and males.

Gynandromorphs in Drosophila and ratio theory of sex determination   In *Drosophila*, occasionally flies are obtained in which a part of the body exhibits female characters and the other part exhibits male characters. Such flies are known as gynandromorphs (Figure 9.6). These are formed due to mis-division of chromosomes and start as female with 2A+2X chromosome but one of the X chromosomes is lost during the division of the cell with the result that one of the daughter cells possesses

2A+2X chromosomes and the other 2A+X. If this event happens during first zygotic division, two blastomeres with unequal number of X chromosomes are found. The blastomere with 2A+2X chromosomes develops into female half, while the second blastomere with 2A+X chromosomes produces male half and the resultant fly is a bilateral gynandromorph. The occurrence of gynandromorphs clearly indicates that the number of X chromosomes determine the sex of the individual.

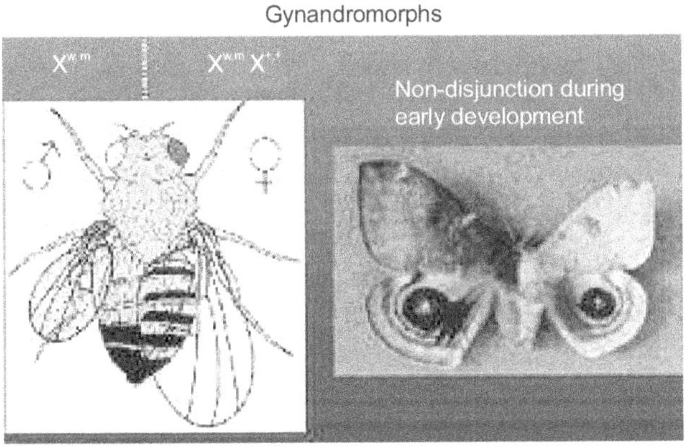

Gynandromorphs

Non-disjunction during early development

Figure 9.6   Gynandromorph in *Drosophila* and moth
(*Source*: biology200.gsu.edu)

HORMONES DETERMINING SEX

Chromosomal theory of sex determination and genic balance theory apply to the lower animals but in higher vertebrates, the embryo develops some characters of the opposite sex together with the characters of its own sex chromosome. This is due to the hormones secreted by the reproductive organs of that animal.

A large number of cases are known where sex is modified due to hormones secreted from sex organs. Some examples are given below.

Sex Reversal

A case of complete sex reversal was reported in 1923 by **Crew**. Crew found in his farm a hen which laid fertile eggs, accidentally lost its ovary, stopped laying eggs, and developed male comb, male plumage and started crowing like a cock. It finally functioned as male and became father of two chickens. Crew's "crowing hen" explains sex reversal due to hormonal control.

This case of sex reversal was explained by assuming that when the ovary is destroyed on removal, the ovarian hormones were stopped. In the place of ovary, the rudimentary gonad (this is a non-functional gonad in birds) started functioning as testis. The male hormones were produced and resulted in the appearance of male secondary sexual characters and the formation of sperms.

The female characters can be induced in a castrated male by injecting female hormones and vice versa.

Freemartin

F.R. Lillie found that in cattles, where twin calves of opposite sex (one male and other female) are born, the male is normal but the female is sterile with many male characteristics. Such sterile females are known as freemartin.

Freemartin is the effect of hormones of the male sex on the female. In cattle, the foetal membranes of the twins are fused in such a manner that they have a common circulation of blood. The female hormone is produced at a slightly later stage in the development and guides its development towards female side. But since the twins have a common circulation and blood passes from one twin into the body of the other twin, the male hormone, which is produced slightly in advance of female hormones, enters the body of female twin and before the onset of development of female characteristics by the female hormone, it is already differentiated in the guidance of male hormone. As a result the developing female is sterile.

Intersexes

The classical work of Goldschmidt on diploid intersexes in gypsy moth, *Lymantria,* is an important contribution towards understanding the mechanism of sex determination.

When *Lymantria dispar* (European race) males are crossed with normal *L. japonica* females, the males and females of the progeny are all normal, but when the sexes were changed, i.e., when females of *L. dispar* of European race were crossed with males of *L. japonica* race, the females were all intersexes, whereas males were all normal.

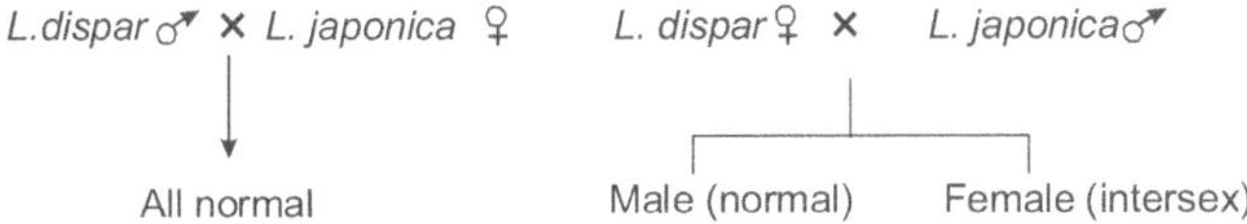

Goldschmidt presumed that the factors for maleness are present on the X chromosome and for femaleness on Y chromosome or in the cytoplasm, and that these factors vary in strength in different races. The sex factors are weak in European strain and strong in Japanese strain. Accordingly, if a European female (XY) is crossed to *L. japonica* male (XX), the females produced are intersexes because X chromosome from *L. japonica* race with its maleness is not completely dominated by the relatively weak femaleness (Y) from European females. On the contrary males receiving one strong X from Japanese race and other weak X from European race result in the formation of normal males.

In the reciprocal cross between *L. japonica* female (strong) and *L. dispar* male (weak) 50% zygotes develop into normal females with weak X and strong Y and the other half into normal males with one strong and one weak X chromosome.

Goldschmidt explained the mechanism as follows. The intersexes start development as females or males in the guidance of their chromosomal configuration and develop as such up to a certain critical point or turning point, after which their development is switched on towards the opposite sex due to the appearance of hormones which direct the development towards that direction. He has postulated certain hypothetical enzymes termed as andrase for initiation of maleness and gynase for femaleness.

ENVIRONMENTAL FACTORS IN SEX DETERMINATION

In a number of organisms, sex is determined fully or partly by environmental factors. One fascinating example is the marine mollusc *Crepidula fornicata* also known as the common slipper limpet. They live in stacks, one on top of the other. Each limpet begins life as a swimming larva. If the larva settles on a solid substrate, it develops into a female. It then produces chemicals that attract other larvae, which settle on top of it. These larvae develop into males which then serve as mates for the limpets below. After a period of time, the males on top develop into females and, in turn attract additional larvae that settle on top of the stack, develop into males, and serve as mates for the limpets under them. The uppermost animals are always male. This type of sexual development is called sequential hermaphroditism; each individual can be both male and female, although not at the same time. In *Crepidula fornicata*, sex is determined environmentally by the limpet's position in the stack.

The most interesting case is the differentiation of sex in *Bonellia viridis* (Figure 9.7). In this worm the larvae are potentially hermaphroditic. But if a single worm is reared from the egg in isolation, i.e., away from other worms of the same species, it invariably develops into a female. On the other hand, if newly hatched worms are released in water containing mature females, young worms which attach to the substratum develop into the females, while others which attach themselves to the proboscis of females differentiate into males which migrate down to the nephridium of females and stay as parasites. Apparently female produces a substance that causes the attached larva to develop into male.

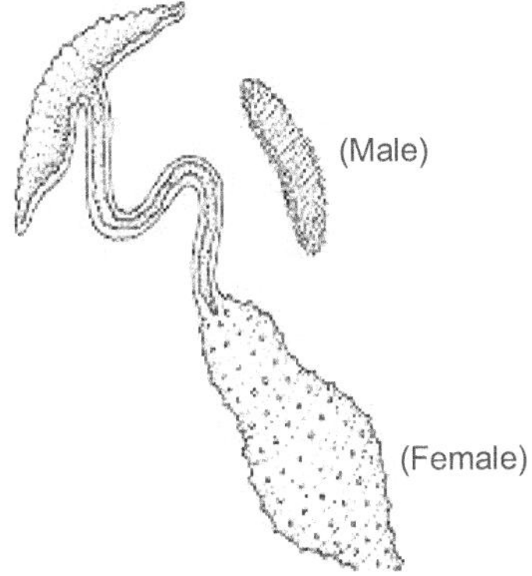

Figure 9.7   *Bonellia*
(*Source*: www.tereska.blog.pl)

Environmental factors are also important in determining sex in many reptiles. Although most snakes and lizards have sex chromosomes, in many turtles, crocodiles and alligators, temperature during embryonic development determines sex. In turtles, for example, warm temperature produces females during certain times of the year, whereas cool temperature produces males. In alligators the reverse is true.

SEX DETERMINATION IN HUMAN BEINGS

Human embryologists had discovered that during the first month of embryonic development, the gonads that develop are neither testes nor ovaries. At about six or seven weeks of development, gonads become either ovaries or testes. For a long time, it was thought that a single gene, a testis-determining factor (*TDF*) located on the Y chromosome acts as a sex switch to initiate male development.

Later it was found that males had a protein on their cell surfaces not found in females; this protein was called the histocompatability Y antigen (H-Y antigen). The gene for this protein was found on the Y chromosome. This acted as a sex-switch, when present direct the gonad to develop into testes and when absent into ovaries.

In 1991, Robin Lovell–Badge and Peter Goodfellow isolated a gene called sex-determining region Y (*SRY*) lying adjacent to *ZFY* gene (zinc finger on the Y chromosome). SRY has been positively identified as the testis-determining factor (TDF) because, when injected into normal (XX) female mice, it caused them to develop as males.

The Sry protein appears to bind at least two genes. One, the p450 aromatase gene, has a protein product that converts the male hormone testosterone to the female hormone estradiol; the Sry protein inhibits the production of p450 aromatase. The second gene that the Sry protein affects is the gene for Mullerian-inhibiting substance, which induces testicular development and the digression of female reproductive ducts; the Sry protein enhances this gene's activity.

Eva Eicher and Linda Washburn have developed a model in which two pathways of coordinated gene action help to determine sex, one pathway for each sex. The first gene in the ovary-determining pathway is termed as ovary determining (*od*) (Figure 9.8). The first gene in the testis-determining pathway must function before the *od* gene begins, in order to allow XY individuals to develop as males. Once the steps of a pathway are initiated, the other pathway is inhibited.

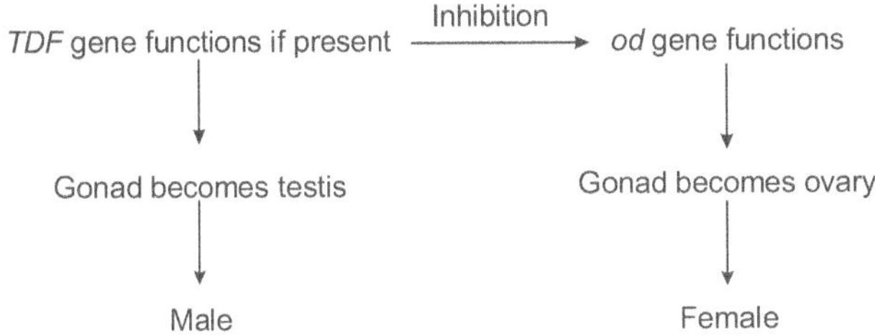

Figure 9.8   A model for the initiation of gonad determination in mammals

Androgen-insensitivity Syndrome

Several genes in Y, in addition to *SRY*, influence sexual development in humans, as illustrated by women with androgen-insensitivity syndrome. These persons have female external sexual characteristics and psychological orientation. In them the vagina ends blindly and the ovary, oviduct and uterus are absent. They fail to menstruate at puberty. Inside the abdomen lies a pair of testes, which produce the male hormone, testosterone. These persons have X and Y chromosome in their cells.

This condition is complex since persons with XY have female appearance. This is due to the complex interaction between genes and testosterones. In these persons, though the male hormones are produced, the cells are insensitive to them and female characteristics develop. It shows that in humans, genes for both sexes are present and the key to maleness or femaleness lies in the control of gene expression.

DOSAGE COMPENSATION

In the XY chromosomal system of sex determination, males have only one X chromosome, whereas females have two. So males have half the number of X-linked alleles as females for genes which are not related to sex traits. In human beings and other mammals, the necessary dosage compensation is accomplished by the inactivation of one of the X chromosomes in females so that both males and females have only one functional X chromosome per cell.

In 1949, M. Barr and E. Bertram first observed a condensed body in the nucleus of a normal female cat and none in a male. They referred the body as Barr body or sex chromatin (Figure 9.9). Mary Lyon then suggested that this Barr body represented an inactive X chromosome, which in females becomes tightly coiled into heterochromatin. This is called Lyon hypothesis. The number of Barr bodies is directly proportional to the extra X chromosomes. XO females have no Barr body. XX females have one and XXXX females have three. Females are considered as sex chromatin-positive and the males are called sex chromatin-negative.

In certain types of blood cells such as the polymorphonuclear leucocytes, the sex chromatin sticks out of the nucleus and consists of a head 1.5 μ in diameter, connected to the nucleus by a filament called drumstick. The frequency of detectable drumsticks is low; 1 in 38 cells against 35–90% for Barr bodies in other tissues.

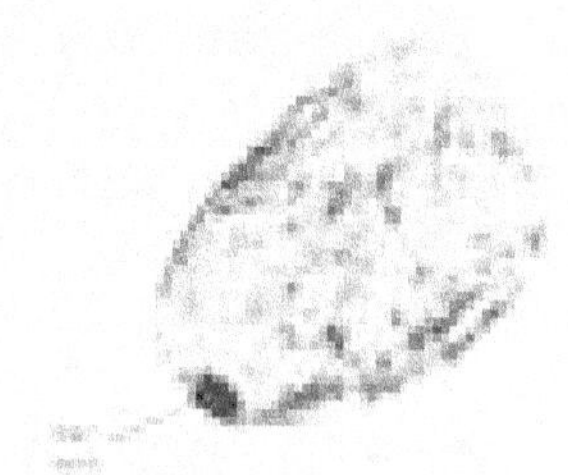

Figure 9.9   Barr body
(*Source*: www.britannica.com)

Evidence for the Lyon Hypothesis

Genetic evidence supports Lyon hypothesis. Females heterozygous for a locus on the X chromosome show a unique pattern of phenotypic expression. In human females, one X chromosome is inactivated randomly in a cell. The same X remains a Barr body for all future generations. Thus heterozygous females show mosaicism at the cellular level for X-linked traits.

Glucose 6-phosphate dehydrogenase (G6PD) is an enzyme controlled by a gene in the X chromosome. The enzyme occurs in two forms, A and B. Since they differ by an amino acid they can be distinguished by electrophoresis. The serum of a female heterozygote has both A and B forms, whereas any single cell has only one or the other, indicating that only one X is active in any particular cell.

The colour phenotype in some mammals is another example. For example, the tortoiseshell pattern of cats is due to the inactivation of X chromosomes. Tortoiseshell cats are normally female heterozygous for the yellow and black alleles of the X-linked colour locus. They exhibit patches of these two colours, indicating that at a certain stage in development, one or the other of the X chromosomes was inactivated and all of the daughter cells in that line kept the same X chromosome inactive. The result is patches of coat colour.

The X chromosome is inactivated starting at a point called the X inactivation centre (XIC). This region contains a gene called Xist (X inactive-specific transcripts). The Xist gene has been identified as the gene that initiated inactivation of the X chromosome. This gene is active only in the inactive X chromosome in a normal XX female. The gene product of Xist is an RNA that is not translated into a protein. It is found that this RNA is associated with Barr bodies, coating the inactive chromosome.

Dosage Compensation in *Drosophila*

Dosage compensation also occurs in fruit flies, but the mechanism is different from that in mammals since no Barr bodies are found in fruit flies. Instead, the male's single X chromosome is hyperactive, reaching the level of activity of both of the female X chromosomes combined. Researchers have discovered a multi-subunit protein complex called MSL (for male-specific lethal) that binds to hundreds of sites on the single X chromosome in males. The binding mediates the hyperactivity of the genes on the X chromosome. At least five genes contribute products to this protein complex: *msl*1, *msl*2, *msl*3, *mle* and *mof*. Along with this protein complex are RNAs that also bind to the male chromosome. These RNAs, also implicated in dosage compensation, are the products of the *rox*1 and *rox*2 genes. Together, the MSL protein complex and the RNAs comprise a compensasome.

Mutant alleles of the male-specific (*msl*) genes disrupt dosage compensation in males and are lethal. However, they have no effect in females. Expression of at least one of these genes, *msl2,* is repressed by the protein product of the *Sxl* gene. Thus sex determination and dosage compensation are ultimately under the control of the same master switch gene *Sxl*.

SEX-LINKED INHERITANCE

Sex chromosomes (XX–XY) are primarily concerned with the determination of sex but they also carry genes for other body characters. Such traits whose genes are located on the sex chromosomes and follow sex during inheritance are known as sex-linked characters. The genes responsible for these traits are termed sex-linked genes and their mode of inheritance is described as sex-linked inheritance.

SEX-LINKAGE IN *DROSOPHILA*

The first extensive experimental evidence for sex-linkage came in 1910 with the discovery by T.H. Morgan of a white-eyed mutant in *Drosophila*. A gene had undergone a change that resulted in phenotypic alteration. This change expressed itself as white eyes rather than the normal red eyes. This white-eyed male was crossed with red-eyed female, the F_1 flies were all red-eyed, indicating that white eye mutation (w) is recessive to red eye (+). When F_1 flies mate freely, the red and white-eyed flies appeared in the ratio of 3 : 1. But all the white-eyed flies were male. The red-eyed males were equally numerous. The females on the other hand were all red-eyed. The white-eyed females did not appear.

Morgan concluded that the gene for eye colour is located on the X chromosome. The F_1 red-eyed females were heterozygous (+). A white-eyed female could only appear when it is homozygous for *ww* genes, i.e., both its X chromosomes possess the recessive gene for white eye colour. A single gene *w* for white eye expresses itself in a male since males have only one X and the Y lacks the allele for eye colour. Such an allelic condition in the male with reference to sex-linked genes is termed as hemizygous (Figure 9.10).

In a reciprocal cross, when a red-eyed male was crossed to a white-eyed female, F_1 offspring consisted of 50% red-eyed and 50% white-eyed and all the red-eyed offspring were females and all the male offspring were white-eyed. When these F_1 offspring were interbred, their F_2 offspring consisted of red- and white-eyed individuals in equal proportion in both the sexes. In *Drosophila*, therefore, sex-linked traits such as white eye colour follows a criss-cross inheritance. The male transmits the sex-linked traits to his grandsons through his daughters who are carriers. The trait is transmitted from one sex to the same sex through the opposite sex.

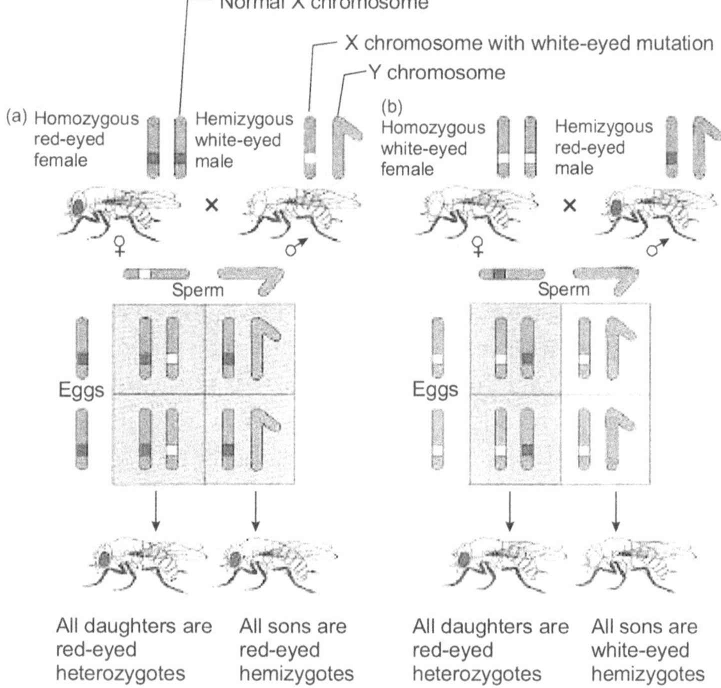

Figure 9.10   Sex-linked inheritance in *Drosophila*
(*Source*: www.fao.org)

SEX LINKAGE IN MAN

Sex linkage has been indicated for more than 200 traits in humans. The most popular examples are (i) haemophilia (ii) colour blindness (iii) myopia (iv) mitral stenosis (v) juvenile glaucoma and (vi) muscular dystrophy.

Red-Green Colour Blindness

Colour perception is controlled by the cones in the retina. The genes for the cone proteins are located in the X chromosome. Red blindness is called protanopia and the green blindness deuteranopia. Colour blindness is caused by X-linked recessive gene.

When a normal woman married a colour-blind man all her sons and daughters have normal colour vision, but when the daughters are married to man with normal colour vision some colour-blind sons are produced.

If a colour-blind woman is married to a normal man, all her sons are colour-blind whereas all the daughters have normal colour vision. When these daughters having normal colour vision are married to a colour-blind man, colour-blind grandsons and granddaughters are produced. It is observed that a colour-blind woman has sons all colour-blind and daughters all with normal vision and a colour-blind woman always has a colour-blind father and her mother is a carrier.

The following conclusions are drawn from the above results.

1.   Colour blindness is more common in males than in females.

2.   Two recessive genes are needed for the expression of colour blindness in female, whereas only one gene gains expression in male.

3.   Males are never carriers.

4.   Colour-blind women always have colour-blind fathers and always produce colour-blind sons.

5.   Colour-blind women produce colour-blind daughters only when their husbands are colour-blind.

6.   Women with normal colour vision, whose fathers are colour-blind, produce normal and colour-blind sons in approximately equal proportion.

7.   The character seems to alternate between the sexes, appearing in females in one generation and in males in the next generation: thus the pattern exhibits criss-cross inheritance.

Haemophilia

Haemophilia is a **bleeder's disease** in humans. In haemophilic man the blood fails to clot when an injury is made resulting in continuous bleeding which can lead to death by loss of blood. Normally the blood clots within 2–8 minutes.

Haemophilia is caused by a sex-linked recessive gene located in the X chromosome (Figure 9.11). It is inherited

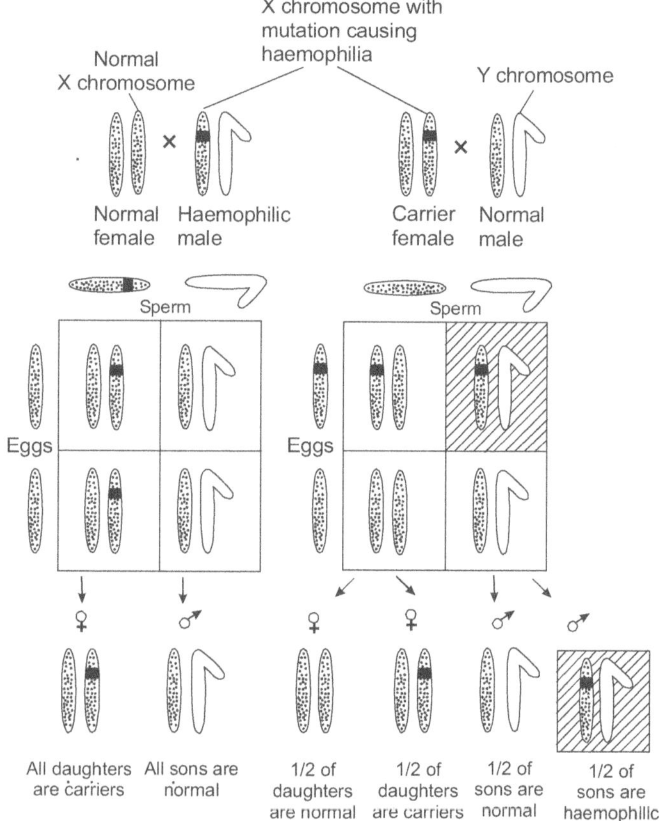

Figure 9.11   Inheritance of haemophilia in man
(*Source*: www.accessexcellence.org)

like colour blindness. A single gene results in haemophilia in man and a woman needs two such genes for the same since she has two X chromosomes and the gene for haemophilia is recessive. Haemophilic men produce daughters who are heterozygous for the haemophilic gene and called carriers and are non-haemophilic. A haemophilic woman could be produced only if a carrier woman of F_1 generation married a haemophilic man.

Haemophilia was observed in Queen Victoria's family and in the Russian Royal families. Haemophilia is a rare genetic disorder. Frequency of haemophilic males is one in 10,000 while that of haemophilic females is one in 10,000,000.

Dominant Sex-linked Genes in Man

Dominant sex-linked diseases are expressed more in women since there are two X chromosomes in woman. The widely studied example is defective enamel of the teeth and is more frequently present in women.

Y-linked Inheritance

The genes in Y chromosome in man are exclusively transmitted through the male line. These traits are also called holandric traits, the genes, holandric genes and the inheritance as holandric inheritance. The gene for hairy pinna is an example for Y-linked genes. The sex-determining genes such as *H–Y* and *TDF* are located in Y chromosomes. 17 genes are found to be present in the Y chromosome of man.

SEX-LINKED GENES IN POULTRY

In poultry, female individual is heterogametic having only one X chromosome and male is homogametic having two X chromosomes. Therefore the inheritance pattern in this case will be reversed.

Barred plumage is a popular example in poultry where the feathers will be banded with bars of black on a white background. The gene for barred plumage is dominant over non-barred plumage. Experimental evidences indicate that a male carries two genes for barred plumage but a female carries only one (Figure 9.12). This is because in birds, males have two X and females have only one.

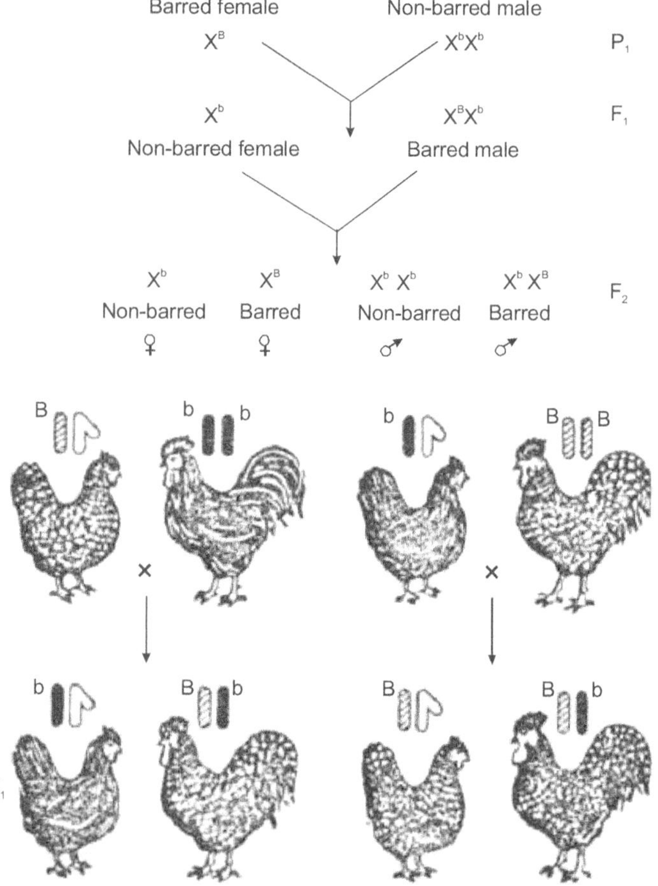

Figure 9.12   Sex-linkage in poultry
(*Source*: www.biology.edu)

The cross between a barred cock and non-barred hen produces only barred offspring. When these offspring interbreed, F_2 generation includes (i) males all barred (ii) 50% of females barred and 50% non-barred.

A reciprocal cross between barred hen and non-barred cock produces males all barred and females all non-barred in F_1 generation and in F_2 both male and female possess barred and non-barred in equal proportions.

SEX-LIMITED INHERITANCE

Sex-limited genes are present in the autosomes but their expression is determined by the presence or absence of one of the sex hormones. Therefore, these genes express only in one sex. These differ from the sex-linked genes which are located in the sex chromosomes. Many of the secondary sexual characters in man depend on sex-limited genes. For example, beard development in human beings is a sex-limited character. The genes for deep male voice and male musculature are expressed only in the presence of male sex hormone. Similarly breast development, genes for feminine voice and feminine musculature are sex-limited characters. Though, both man and woman have genes for beard and mammary gland development, due to sex hormones the female lacks beard and male lacks developed breasts.

Milk production in cattle is also sex-limited, since genes for milk production are carried by both males and females, but they express only in females. Another example is plumage pattern or feathering in birds. Many birds have a marked sexual dimorphism in feathering pattern. In domestic leghorn fowl, males have long, pointed, curved, fringed feathers on tail and neck, but feathers of female are shorter, rounded, straighter and without the fringe. Thus males are cock-feathered and females are hen-feathered. It has been shown that hen feathering results from a single gene H, and cock-feathering from its allele h as follows.

Table 9.2   Sex-limited genes in birds

Genotypes	Phenotypes	
	Males	**Females**
HH	Hen-feathering	Hen-feathering
Hh	Hen-feathering	Hen-feathering
hh	Cock-feathering	Hen-feathering

It has been found that the expression of gene H and h depends upon the sex hormones because when the ovaries or testes of chicken have been removed (gonadectomized), all become cock-feathered at next moult following surgery, regardless of genotype (Table 9.2). It shows that particular type of feathering depends upon specific combination of genotype and sex hormones, and the H gene produces hen-feathering in the presence of either sex hormones, and cock-feathering in the absence of any hormone. The 'h' gene produces cock-feathering if female hormone is absent and hen-feathering if female hormone is present.

SEX-INFLUENCED INHERITANCE

Sex-influenced traits are determined by autosomal genes and are inherited according to Mendel's principles, but they are expressed differently in males and females.

For example, the presence of beard in some goats is determined by an autosomal gene B that is dominant in males and recessive in females (Figure 9.13). In males a single allele is required to express the trait: both the homozygotes BB and heterozygotes Bb have beards, whereas bb male is beardless.

In contrast, females require two alleles in order to be bearded. The homozygote female BB is bearded, whereas the heterozygote Bb and recessive bb are beardless (Table 9.3).

Figure 9.13   Bearded goat
(*Source*: donowdo.com)

Table 9.3   Sex-influenced genes in goats

Genotypes	Phenotypes	
	Males	Females
BB	Bearded	Bearded
Bb	Bearded	Beardless
bb	Beardless	Beardless

An example for sex-influenced trait in man is pattern baldness, in which hair is lost prematurely from the front and top of the head and so only a fringe of hair remains. Pattern baldness is autosomal. Men require only a single bald allele to become bald, whereas women require two alleles, and so baldness is common among men. The expression of the allele for pattern baldness is clearly enhanced by the presence of male sex hormones.

Summary

- ✿ The mechanism by which sex is established is termed as sex determination.

- ✿ There are various mechanisms operating in determining the sex of different organisms. There are four major kinds of sex-determination mechanisms. They are chromosomal sex-determining systems, genic balance system, hormonal system and environmental factors.

- ✿ In chromosomal mechanism, one of the sex is homogametic and the other sex is heterogametic.

- ✿ Bridges worked out the ratio theory of sex-determination in *Drosophila.*

- ✿ According to this theory the ratio of X chromosomes to autosomes is the determining factor for sex. Sex index is calculated to find out the sex in *Drosophila.* In *Drosophila* a sex-switch gene has been discovered that directs female development.

- ✿ In *Drosophila,* occasionally flies are obtained in which a part of the body exhibits female characters and the other part exhibits male characters. Such flies are known as gynandromorphs. These are formed due to mis-division of chromosomes.

- ✿ A large number of cases are known where sex is modified due to hormones secreted from sex organs. Sex reversal in fowl and occurrence of free martin are the examples. In a number of organisms, sex is determined fully or partly by environmental factors as shown in *Crepidula* and *Bonellia.*

- ✿ In human beings and other mammals, the necessary dosage compensation is accomplished by the inactivation of one of the X chromosomes in females so that both males and females have only one functional X chromosome per cell. The phenomenon is explained by Lyon hypothesis.

- ✿ The genes responsible for some traits are present in sex chromosomes and termed sex-linked genes, and their mode of inheritance is described as sex-linked inheritance.

✿ White eye in *Drosophila* is the example for sex linkage and sex linkage has been indicated for more than 200 traits in humans. The most popular examples are haemophilia and colour blindness. The genes in Y chromosome in man are exclusively transmitted through the male line. These traits are also called holandric traits.

✿ Sex-limited genes are present in the autosomes but their expression is determined by the presence or absence of one of the sex hormones. Therefore, these genes express only in one sex.

✿ Sex-influenced traits are determined by autosomal genes and are inherited according to Mendel's principles, but they are expressed differently in males and females.

REVIEW QUESTIONS

1. What is the difference between female-determining and male-determining sperms in heterogametic males? What influence does each type have in determining sex?

2. In line with Bridges' genic balance theory, what is the expected sex of individuals with each of the following chromosome arrangements?

   i.   4X4A

   ii. 3X4A

   iii. 2X3A

   iv. 1X2A

3. A male with a single Y chromosome is found to have a Barr body. What is his sex-chromosomal constitution?

4. How would you demonstrate Lyon hypothesis?

5. What is a sex switch? What genes serve as sex switches in man and *Drosophila*?

6. Colour blindness in humans is due to a sex-linked recessive allele. Amritha has normal vision, but her mother is colour-blind. Amritha marries Kumar who is colour-blind. What is the probability that their children will be colour-blind?

7. Ramesh has classic haemophilia which is an X-linked recessive trait. From whom could have Ramesh inherited the disease?

    i. maternal grandmother

    ii. maternal grandfather

    iii. paternal grandmother

    iv. paternal grandfather

    Justify your conclusion.

8. Differentiate sex-limited genes from sex-influenced genes with an example.

10

GENE MUTATIONS AND CHROMOSOMAL VARIATION

Objectives

- ☼ To analyse point mutations and molecular mechanisms involved
- ☼ To learn the kinds of mutagenic agents and their role in mutagenesis
- ☼ To understand methods of mutation detection
- ☼ To know the kinds of chromosomal mutations and their impact

Key Terms

mutation	Ancon	trinucleotide repeats
chromosomal aberrations	frameshift mutation	pseudodominance
mutant	cri-du-chat syndrome	point mutation
mutation rate	shift translocation	mutable genes
mutator genes	paracentric inversion	anti-mutations
deletion	substitution mutation	transition
copy error mutation	tautomerization	ionization
base analogs	5-bromouracil	aminopurine
deamination	depurination	transversion
CIB method	Ames test	mutagens
target theory	anoxia	photoproducts
dimers	photoreactivation	simple translocation
intercalary	buckling effect	notch character
bar eye	aneuploidy	doubling
euploidy	trisomic	inversion

position effect	alternate segregation	monosomy
pericentric inversion	nullsomy	forbidden base pairs
tetrasomy	trisomy 13	terminal deletion
familial Down syndrome	trisomy 18	Edward syndrome
Patau syndrome	trisomy 8	gynecomastia
reciprocal translocation	turner syndrome	gonadal dysgenesis
mosaicism	chimeras	Klinefelter syndrome
autopolyploidy	gynandromorph	allopolyploidy
displaced duplication	tandem duplication	suppressor mutation
multiple translocation	amphidiploid	
reverse tandem duplication		
deletion or interstitial deletion		

INTRODUCTION

DNA is a highly stable molecule that replicates with amazing accuracy, but changes in DNA structure and errors of replication do occur. Hugo de Vries used the term mutation to describe heritable phenotypic changes in organisms. A mutation is defined as an inherited change in genetic information. Mutations are both the sustainer of life and the cause of great suffering. On the one hand, mutation is the source of all genetic variations that are inevitable for biological diversity and evolution as they are the raw materials for evolution. On the other hand, most mutations have detrimental effects, and mutation is the source of many human diseases.

The earliest record of point mutations dates back to 1791, when Seth Wright observed a short-legged breed of sheep caused by mutation. This breed was called Ancon breed and was preferred because they could not get over the fence and damage the crop in the adjacent field. In 1910, T.H.Morgan reported a white-eyed mutant in *Drosophila melanogaster*.

At present about 500 different mutations are observed in fruit flies.

There are two broad categories of mutations: somatic mutations and germ-line mutations.

Somatic mutations originate in somatic tissues. These cells do not produce gametes. These mutations are passed on to other cells through the process of mitosis. Many somatic mutations are not harmful. But a cell with somatic mutation can undergo rapid cell division and thus increase in number, and this is the basis for all cancers.

Germ-line mutations arise in cells that produce gametes. These mutations can be passed on to future generations.

Mutations have been also categorized as gene mutations, that affect a single gene, and those that affect the number or structure of chromosomes called chromosomal mutations or chromosomal aberrations.

GENE MUTATIONS

Inheritance is based on genes that are faithfully transmitted from parents to offspring during reproduction. Mechanisms have evolved to facilitate the faithful transmission of genetic information from generation to generation. Nevertheless mistakes or changes in the genetic material do occur. Such sudden, heritable changes in the genetic material are called mutations. An organism exhibiting a novel phenotype as a result of the presence of a mutation is referred to as a mutant.

The term mutation was introduced by Hugo de Vries. The mutation which consists of single changes in the nucleotide sequence is termed as point mutation. If the change is a replacement of some kind, then a new codon is created. In many cases, new proteins appear. These new proteins can alter the morphology or physiology of the organism and result in a new phenotype.

Mutation Rate

The mutation rate is the number of mutations that occur per cell division in bacteria and single-celled organisms, or the number of mutations that arise per gamete in higher organisms. For *E. coli* the mutation rate is 10^{-5} to 10^{-7}, and varies for different genes. The lethal mutation rate in *Drosophila* is about 1×10^{-2} per gamete. Genes associated with human traits such as intestinal polyposis and muscular dystrophy have been estimated to mutate once in 10^4 to 10^5 people.

Some genes mutate more frequently than others. Such genes are called mutable genes. The genes that influence the mutation rate of other genes are called mutator genes. Some mutator genes reduce the frequency of mutation of certain genes and are known as **antimutators** or suppressor mutators.

MOLECULAR BASIS OF GENE MUTATION

Though very accurate, the DNA replication sometimes shows inaccuracy which may occur within the DNA due to external or internal, natural or artificial factors. These introduce changes in the polynucleotide chain of a DNA molecule. Point mutations may be of the following types.

1. Substitution mutation
2. Frameshift mutation

Substitution Mutation

In a substitution mutation, a nitrogenous base of a triplet codon of DNA is replaced by another nitrogen base, changing the codon. The altered codon may code for a different amino acid and may result in the formation of a protein molecule with a single amino acid substitution resulting in an altered phenotype. Substitution may be of two types: transition and transversion.

Transition   These are changes that involve replacement of one purine in a polynucleotide chain by another purine and correspondingly in the complementary chain the replacement of one pyrimidine by another pyrimidine. The transitional substitutions can be introduced by any of the following ways either during DNA replication (copy error mutation) or otherwise.

i. Tautomerization   In a normal DNA, the purine, adenine (A), is linked to the pyrimidine, thymine (T), by two bonds, while the purine, guanine (G), is linked to pyrimidine, cytosine (C), by three bonds. However all these nitrogenous bases exist in alternate states. These states are called tautomers and are formed by the rearrangements in the distribution of hydrogen atoms (tautomeric shifts). Due to tautomerization, the amino ($-NH_2$) group of cytosine and adenine is converted to imino (–NH) group and likewise keto group ($C=O$) of thymine and guanine is converted to enol group (–OH).

In tautomeric state, a nitrogenous base cannot pair with its normal partner. A tautomeric adenine pairs with normal cytosine and tautomeric guanine with thymine. Similarly tautomeric thymine pairs with normal guanine and tautomeric cytosine with adenine. Such pairs of nitrogenous bases are known as forbidden base pairs or unusual base pairs.

The rare bases can introduce mutations during DNA replication. For example, adenine in the parent DNA is in rare state, the complementary new chain will contain cytosine. At the time of next replication this cytosine would pair with guanine. This will produce a substitution of A=T base pair by

$G \equiv C$ pair. Similarly a situation of $G \equiv C$ by A=T pair can be produced if cytosine is in tautomeric state (Figure 10.1).

ii. Ionization   Transitions may also be introduced by ionization of a base at the time of DNA replication. Ionization involves the loss of the hydrogen from 1st position of nitrogen in a nitrogenous base. For example, in its ionized state, thymine pairs with normal guanine and ionized guanine links with normal thymine.

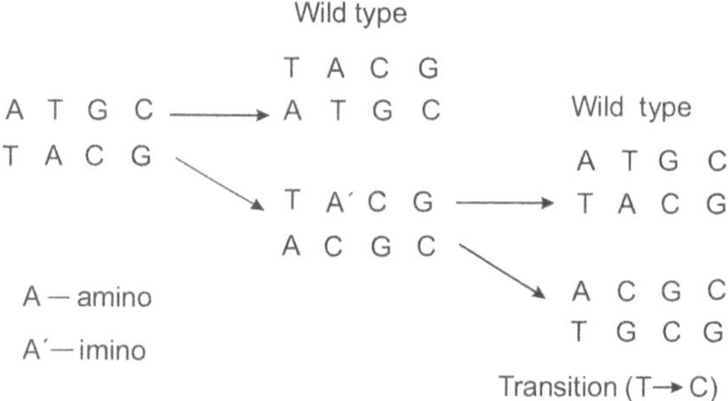

Figure 10.1   Transition mutation
(*Source*: www.mun.ca)

iii. Base analogs   Certain chemical compounds have molecular structure similar to the nitrogenous bases present in DNA nucleotides. These are called **base analogs**. These are usually derivatives of nitrogenous bases of DNA and occur as natural or artificial base analogs. Some of the natural base analogs are 5-methylcytosine, 5-hydroxymethylcytosine, 5-glucosyl hydroxymethylcytosine, 5-hydroxymethyl uracil and 6-methylpurine. The artificial base analogs include 5-bromouracil (5-BU), 5-iodouracil (5-IU), 2-bromo and 5-methylcytosine. The former two are base analogs of thymine and latter those of cytosine.

5-Bromouracil   It is a structural analog of thymine. Its chemical structure is very much like that of thymine. Its keto form is more common while, the enol state is rare or it may occur in ionized state. Its normal keto form pairs with adenine and rare enol form pairs with guanine (Figure 10.2). When such mispairing (G=BU) occurs and DNA undergoes further replication, G pairs with C, while keto BU pairs with A which on further replication pairs with T. Thus $G \equiv C$ pair is replaced by $A = T$ base pair.

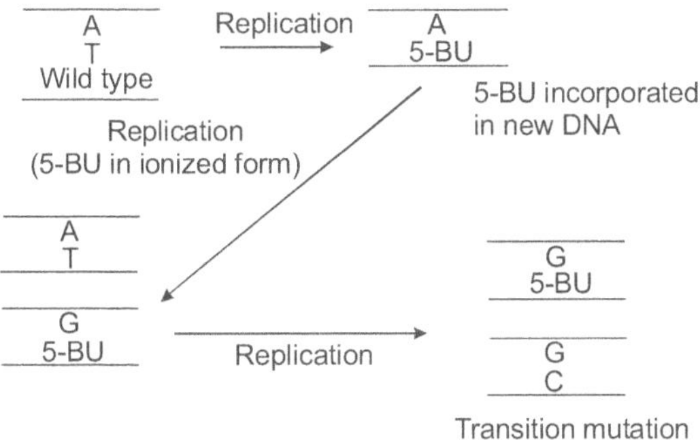

Figure 10.2   Copy error mutation
(*Source*: www.answers.com)

Aminopurine   It is an artificial chemical base analog of adenine. It can pair with thymine by two hydrogen bonds and with cytosine by a single bond. Incorporation of aminopurine (AP) in the place of guanine to give AP-C base pair will cause mutation in subsequent generations. Similarly a mistake in replication after incorporation of aminopurine, leading to the formation of AP-T base pair may lead to mutation.

iv. Deamination   Certain chemical substances like nitrous acid, hydroxylamine, diethyl sulphate (DES), ethyl methane sulphate (EMS), ethyl ethane sulphate (EES), nitrosoguanidine (NTG) and nitrosomethyl urea (NMU) change the base sequence in DNA by a series of chemical steps. Some of them like nitrous acid and hydroxylamine cause deamination of nitrogenous bases by replacing amino group by hydroxyl group. The deamination of cytosine leads to the formation of uracil, deamination of adenine forms hypoxanthine (H) and that of guanine forms xanthine. Hypoxanthine exhibits bonding similarity with guanine.

At the time of DNA replication, uracil pairs with adenine and hypoxanthine pairs with cytosine. This leads to the substitution of A = T for G $\equiv$ C and G $\equiv$ C for A = T.

Alkylating agents like DES, EES and EMS may cause depurination (Figure 10.3). Alkylation of guanine at the 7th position gives rise to quaternary nitrogen which is unstable. This either hydrolyses the alkyl group or alkylated purine separates from deoxyribose sugar leaving it depurinated. This gap may then interfere with DNA duplication or may cause incorporation of wrong base. The depurinated DNA is also more labile and may undergo breakage in the backbone soon after. This may induce large alterations in DNA.

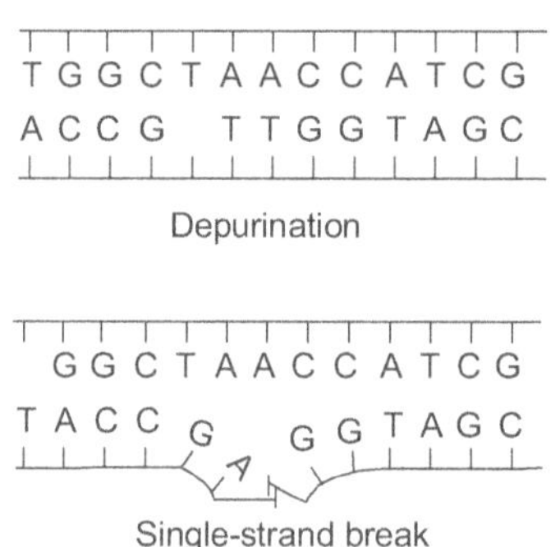

Figure 10.3   Effect of alkylating agents
(*Source*: www.answers.com)

v. Transition caused by hydroxylamine   The hydroxylamine causes hydroxylation of cytosine at amino group giving rise to hydroxylcytosine, which then pairs with adenine rather than with guanine. This at the time of DNA replication introduces T = A substitution.

Transversion   Certain alkylating agents like ethyl methane sulphate (EMS) and methyl methane sulphate (MMS) induce substitution by two ways.

i.   By substituting a purine for purine or pyrimidine for a pyrimidine (transition).

ii.   By substituting purine for pyrimidine or a pyrimidine for a purine (transversion), i.e., A = T to C $\equiv$ G.

For causing transversions, these chemical alkylate the purine nitrogenous bases in the nitrogen at the seventh position in the guanine and adenine and finally lead to its separation from the DNA strand. This is known as **depurination**. Depurination leaves a gap at that point. At the time of replication, any of the four bases can possibly get inserted at this place in the complementary strand. If the inserted nucleotide contains a pyrimidine, it is transition and if purine then it is transversion. In the next cycle of DNA synthesis a DNA molecule is formed which contains the complete transversion.

Reactive forms of oxygen (including superoxide radicals, hydrogen peroxide and hydroxyl radicals) are produced in the course of normal aerobic metabolism, as well as by radiation, ozone, peroxides, and certain drugs. These reactive forms of oxygen damage DNA. For example, oxidation converts guanine into 8-oxy-7,8-dihydrodeoxyguanine, which frequently mispairs with adenine instead of cytosine, causing a GC to TA transverse mutation.

Frameshift Mutations

A point mutation may consist of replacement, addition or deletion of a base (Figure 10.4). This kind of mutation is frameshift mutation. The addition or deletion of a base produces dangerous effects on the cell or organism because they change the reading frame of a gene from the site of mutation onward.

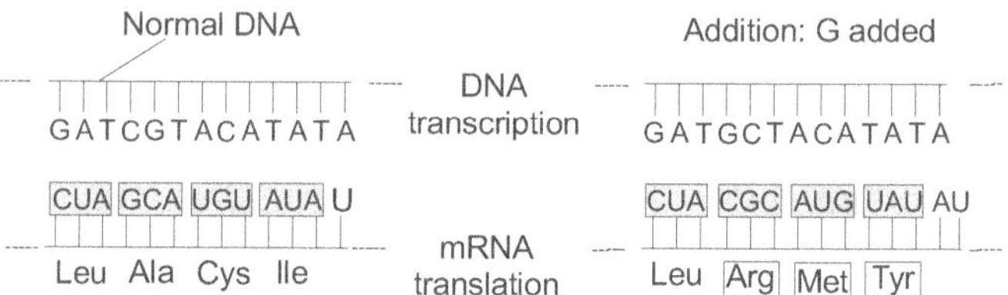

Figure 10.4 Frameshift mutation
(*Source*: www.staff.tushsd.k12.az.us)

A frameshift mutation causes two problems. First, all the codons from the site of change onward will be different and thus yield a useless protein. Second, stop-signal information will be misread. One of the new codons may be a nonsense codon, it is no longer recognized as such because it is in a different reading frame, and therefore, the translation process continues beyond the end of the gene.

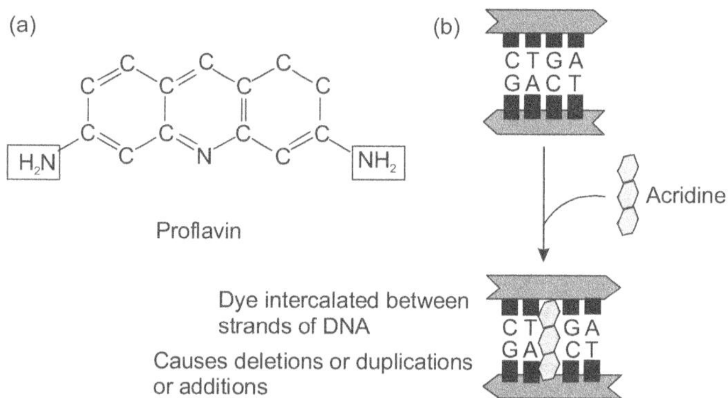

Figure 10.5 Mutation by intercalating agents
(*Source*: ww.academic.brooklyn.cuny.edu)

Origin of frameshift mutation Acridine dyes cause deletion or insertion of a single base pair. Acridines like 5-aminoacridine and proflavin become intercalated between two adjacent purines and thus increase the distance between them from 3.4 Å to 6.8 Å (Figure 10.5). At the time of DNA replication, either a nitrogenous base pair is introduced in the gap or a nitrogenous base pair is lost.

Expanding Trinucleotide Repeats

In 1991, an entirely novel type of mutation was discovered. This mutation occurs in a gene called *FMR*-1 and causes fragile-X syndrome, the most common cause of hereditary mental retardation. The *FMR*-1 gene contains a number of adjacent copies of the trinucleotide CGG. In normal allele about 60 or fewer copies of repeats are seen but in mutated gene hundreds or even thousands of copies of the trinucleotide are observed.

Expanding trinucleotide repeats are seen in other human diseases as well. In Huntington disease the trinucleotide CAG expands. The single-stranded regions of some trinucleotide repeats are known to fold into hairpins and interfere with normal DNA replication.

DETECTION OF MUTATION

The evaluation of genetic risks from exposure to environmental chemicals is quite complex owing to number of variables. There is a growing concern over the mutagenicity and carcinogenicity of chemicals that are already present or are constantly entering our environment. A large number of test systems and methodologies are currently available to screen for mutagenicity of environmental chemicals and these encompass a variety of cell types *in vitro*, from bacteria and phage to human cells as well as tests that can be done in whole animals or human beings.

ClB Method

H.J. Muller devised ClB method (Figure 10.6) for the detection of X-linked mutations in *Drosophila*. Muller made a cross between ClB heterozygous female and an irradiated wild type male to detect the lethals that would have been induced by X-rays in X chromosome of irradiated male.

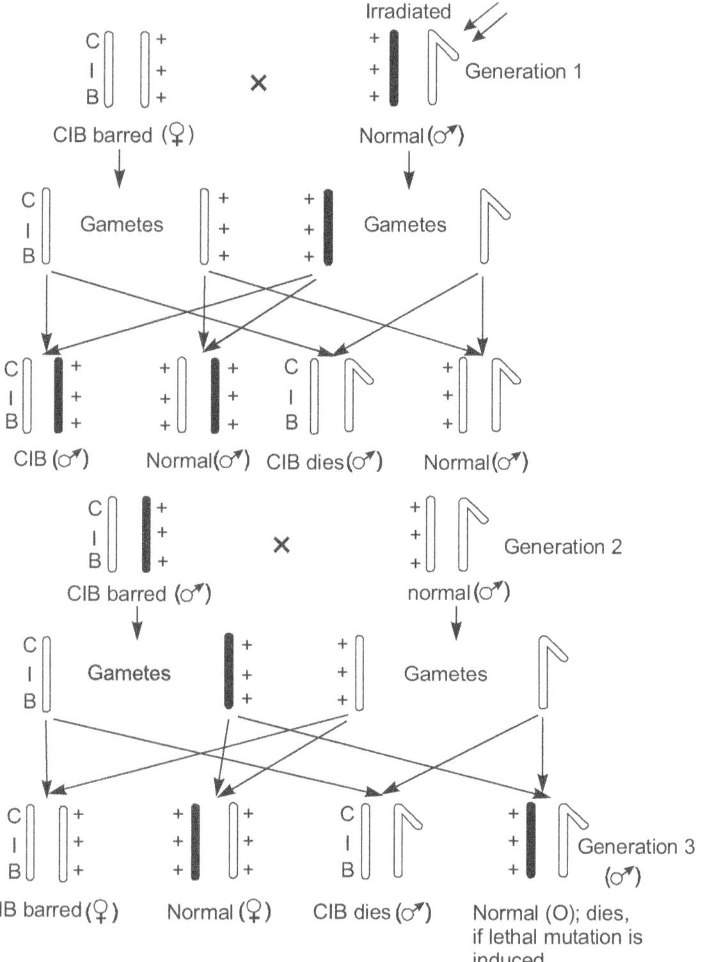

Figure 10.6   Detection of mutation by CIB method (Adapted from King, 1965)

In ClB, B designates a sex-linked dominant mutation producing bar-shaped eyes. It is used as marker for flies.

l stands for a recessive lethal mutant present in the X chromosome. Female *Drosophila* homozygous for this gene l/l and male hemizygous l/y are nonviable. C represents a long inversion that prevents crossing over within the inverted segment. Genes l and B are located in this inverted segment.

The first generation resulting from the sperm of X-rayed male is known as the X_1 generation, since X-rays were used as mutagen. When the X_1 ClB barred females were mated with a wild male, the X_2 generation consisted of a normal and bar-eyed females but no males. All males with the ClB chromosome died because the lethal gene (l) produced its lethal effect in hemizygous condition. Likewise all males with treated X chromosome also die indicating that a lethal mutation has taken place on the X chromosome of an irradiated male parent by X-ray radiation. Therefore, the X_2 progeny consists of only females. If mutation introduced by radiation is not lethal, it would appear among the 50% surviving males of F_2 progeny.

Ames Test

The human population is exposed to a multitude of hazardous chemicals. Some of them are mutagenic and carcinogenic. In 1974, Bruce Ames developed a simple test for evaluating the potential of chemicals to cause mutations. The Ames test is based on the principle that both cancer and mutations result from damage to DNA. The results of experiments have confirmed that 90% of known carcinogens are also mutagens. The mutagenesis in bacteria could serve as an indicator of carcinogenesis in humans.

The Ames test (Figure 10.7) uses four strains of the bacterium *Salmonella typhimurium* that have defects in the lipopolysaccharide coat, which normally protects the bacteria from chemicals in the environment. Furthermore, their DNA repair system has been inactivated, enhancing their susceptibility to mutagens.

One of the four strains used in the Ames test detects base-pair substitutions; the other three detect different types of frameshift mutations. Each strain carries a mutation that renders it unable to synthesize the amino acid histidine (*his⁻*), and the bacteria are plated on to medium that lacks histidine. Only bacteria that have undergone a mutation (*his⁺*) are able to synthesize histidine and grow on the medium. Different dilutions of a chemical to be tested are added to plates inoculated with the bacteria, and the number of mutant bacterial colonies that appear on each plate is compared with the number that appears on control plates with no chemical. Any chemical that significantly increases the number of colonies appearing on a treated plate is mutagenic and is also probably carcinogenic.

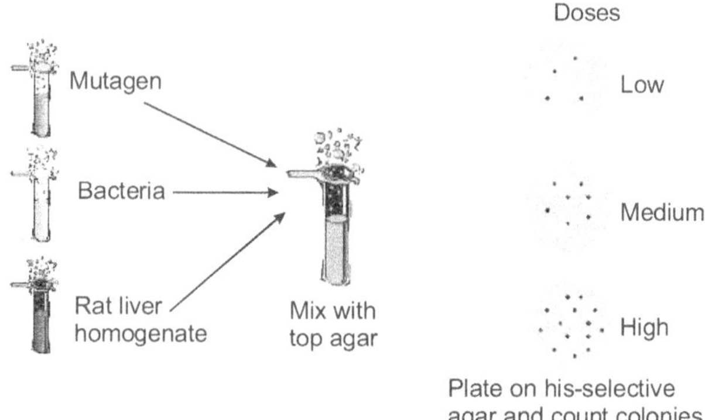

Figure 10.7   Ames test
(*Source*: www.uky.edu)

MUTAGENS

Although many mutations arise spontaneously, a number of environmental agents are capable of damaging DNA, including certain chemicals and radiation. Mutagens are physical, chemical or biological agents that significantly increase the rate of mutation above the spontaneous rate.

Radiation

All forms of energy radiations that are capable of disrupting the chemical structure of genes have been found to be mutagenic. These include UV rays, X-rays, alpha rays, gamma rays and beta rays.

Biological effects

Effect of ionizing radiations   In 1927, H. Muller demonstrated that mutations in fruit flies could be induced by X-rays. Because of their high energy, X-rays, gamma rays and cosmic rays are capable of penetrating tissues and damaging DNA.

Ionizing radiations include X-rays, gamma rays, alpha rays, beta rays, neutrons and protons. The alpha and beta rays do not penetrate beyond the human skin, and therefore, do not affect body cells. The gamma and X-rays collide with the molecules of the cells at high speed and eject electrons from the outer shells of atoms. The atoms therefore, become positively charged. In addition, the ejected electrons move at high speed and, in turn, knock other electrons free from their respective atom. When energy is dissipated, the free electrons attach to other atoms which then become negatively charged ions. These ions then undergo chemical reactions producing mutagenic effects.

These radiations produce breaks in the chromosomes and chromatids and abnormal mitosis in cells. The chromosome breaks lead to loss of chromosome segments, interchanging of segments and loss of genes.

The frequency of mutants per viable organism often increases linearly with the dose. Timofeeff-Ressovsky *et al.*

(1935) interpreted this relationship in terms of a target theory. This theory states that single hit of the particle on the target (DNA) mutates it and it occurs at random. The frequency of single chromosomal break arises by single hit. The frequency of two hit aberrations is more if radiation is given continuously.

Some chemicals are found to protect organisms from radiation and some other chemicals could increase the effect. For example, low oxygen concentration reduced the frequency of chromosome breaks induced by radiations. This is called oxygen effect or anoxia effect.

Radiations in the presence of oxygen from some peroxide radicals may influence the frequency of breaks and mutations. The peroxides cause breaks and also prevent rejoining. Ionization of water in cells may give free radicals and hydrogen peroxide in the following manner.

i. $H_2O \leftrightarrow H^0 + OH^0$ (free radicals)

ii. $H^0 + H^0 \leftrightarrow H_2$

iii. $OH^0 + OH^0 \leftrightarrow H_2O_2$

Effect of non-ionizing radiations   Ultraviolet rays have longer wavelengths and carry much more lower energy. The UV rays are absorbed by nucleic acids and cause alterations in the bond characteristics of purines and pyrimidines.The bases so altered are called photoproducts. Pyrimidines are more prone to such changes. Two adjacent pyrimidines of the same DNA strand are found to form covalent bonds forming dimers. This dimerization (Figure 10.8) interferes with the proper base-pairing of thymine with adenine and may result in the pairing of thymine with guanine. This results in base substitution.

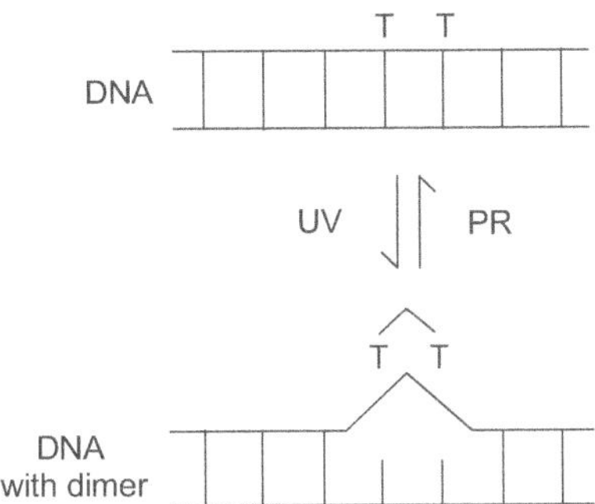

Figure 10.8   Dimerization by UV rays
(*Source*: www.dl.clackamas.edu)

An unusual effect of UV radiation was discovered by **Kelner**, who observed that UV effect may be reversed by exposing the cells to visible light containing wavelengths in the blue region of the spectrum. This phenomenon is known as **photoreactivation** (PR). This has been carried out by an enzyme that splits the thymine dimers and repairs the DNA damage. In human beings, when this repair system is impaired, it causes the skin disease **xeroderma pigmentosum** which causes the patient to be susceptible to sunlight since the enzyme required for removing the thymine dimers produced due to UV is absent.

Chemical Mutagens

The possible sources of chemical mutagens are unlimited. The list of chemical mutagens becomes longer with every passing year. Some of the most powerful mutagens are mustard gas, formaldehyde, ethylurethane, organic peroxides, acridine dyes, nitrous acid, 5-bromouracil and LSD (lyseric acid diethylamide).

CHROMOSOMAL VARIATIONS

Each species of animals or plants has a fixed number of chromosomes. These chromosomes are represented as one set in haploid gametes and twice in diploid body cells. Sometimes changes occur in the structure or number of the chromosomes. These bring in major alterations in the cells or organisms. There are two major chromosomal alterations. They are associated with (i) structure of the chromosomes (ii) number of the chromosomes.

CHROMOSOMAL MUTATIONS

There are visible changes in the structure of chromosomes, involving changes either in the total number of genes or gene loci in a chromosome or their rearrangement. These are also known as chromosomal aberrations. These arise from breaks in chromosomes and may be of the following four types:

Changes involving the number of gene loci:

1.   Deficiency or deletion

2.   Duplication or addition

Changes involving the arrangement of gene loci:

3.   Translocation

4.   Inversion

Deletion

Deletion of a chromosomal segment resulting in the loss of genes is called deficiency (Figure 10.9). Depending on the length of the lost segment, the genes lost may vary from a single gene to a block containing many genes. A chromosome with segments ABCDE ● FGHIJ (● represents the centromere) undergoing deletion of the segment C would generate the mutated chromosome ABDE ● FGHIJ.

The break in the chromosome may be caused by several agents such as chemicals, radiations and viruses. The break occurs at random either in both the chromatids of a chromosome (chromosome break) or only in one chromatid (chromatid break).

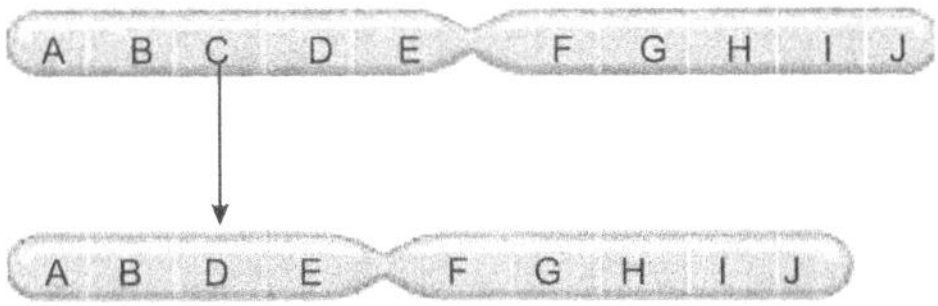

Figure 10.9   Deletion
(*Source*: www.trc.ucdavis.edu)

The segment that is deleted does not survive, since it lacks the centromere. The portion of the chromosome carrying the centromere functions as a genetically deficient chromosome.

Types   Deletion may be of two types (Figure 10.10).

Terminal deletion   It refers to the loss of segment from one or the other end of the chromosome. The terminal part fails to survive and causes terminal deletion. It is caused by a single break in the chromosome.

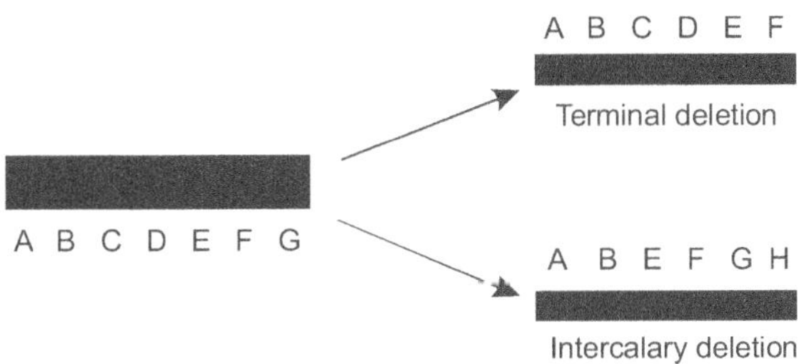

Figure 10.10   Types of deletion

Intercalary deletion or interstitial deletion   It involves two breaks and an intermediate segment is deleted followed by the reunion of the terminal segments.

When an intercalary part of a chromosome is missing, a buckling effect may be observed microscopically, and this buckle-like projection is formed by the normal part of the chromosome which corresponds to the missing part of the deletion chromosome.

Bridges observed deletion in *Drosophila*. A mutant produces notched margin of the wings and is known as notch character, a lethal trait. In humans, a deletion on the short arm of chromosome 5 is responsible for cri-du-chat (cry of the cat) syndrome.

Significance Since deletions involve loss of genetic material, these have some deleterious effect on the organism, which is dependent upon the amount and quality of genetic material lost. A deletion might be small without producing any detectable change in the organism. But the deletions of large size are lethal to the organisms. Deletions may allow recessive mutations on the undeleted chromosome to be expressed. This phenomenon is referred to as **pseudodominance**.

Duplication

The presence of same block of genes more than once in a haploid complement is known as duplication and the additional segment is called a repeat. Consider a chromosome with segments ABC•DEFG. A duplication might include the DE segment (Figure 10.11).

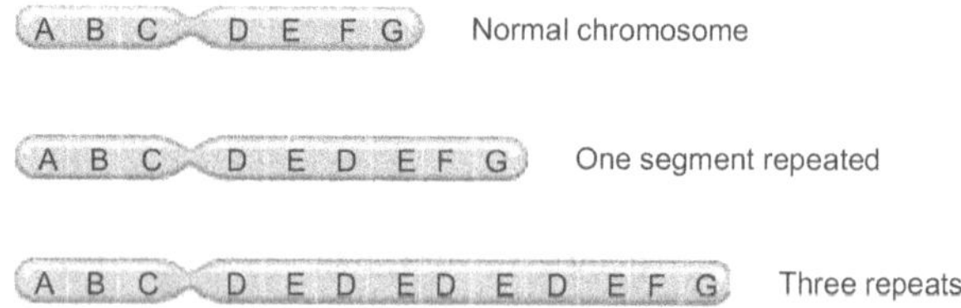

Figure 10.11 Duplication
(*Source*: www.trc.ucdavis.edu)

Types   Three types of duplication have been recognized (Figure 10.12).

Tandem duplication   In this duplication the added segment has the same genetic sequence as is present in the original state in the chromosome. For example, if duplication piece is ABC, a tandem duplication will be ABCABCDEF.

Reverse tandem duplication   In such duplication the sequence of genes aligned in the attached chromosome piece is just the reverse of the original segment. For example, if duplicated piece is ABC, the reverse tandem will be ABCCBADEF.

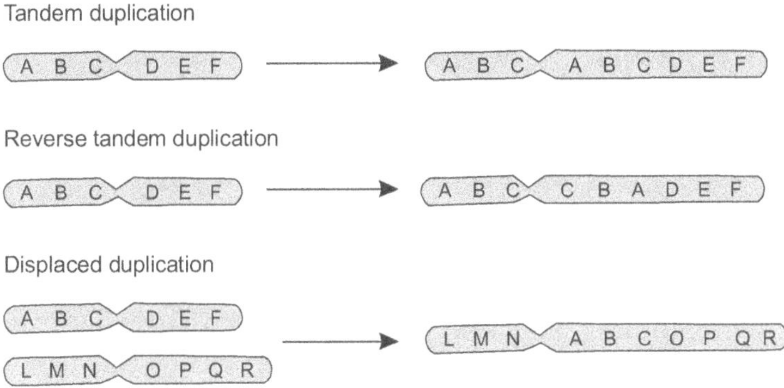

Figure 10.12   Types of duplication

Displaced duplication   In displaced duplication the chromosomal segment gets attached to some nonhomologous chromosome (with segments LMNOPQR), as for example, LMNABCOPQR.

An example of duplication in *Drosophila* is the bar eye. The gene for bar eye is located in the 16A region of the X chromosome. In homozygous bar eye there are two 16A regions. In heterozygous double bar, the 16A is represented three times and the effect is severe.

Significance Duplications are more frequent and less deleterious. These do not lower the viability, but do produce abnormality of structure and function. Duplication plays a significant role in evolution, because duplication increases the number of genes in the chromosome complement.

Translocation

Translocation is a kind of chromosomal rearrangement in which a block of genes from one nonhomologous chromosome is transferred to another nonhomologous chromosome (Figure 10.13). Translocation is different from crossing over, in which there is exchange of genes between homologous chromosomes. If a chromosome has AB genes and another chromosome CD and exchange of genes result in exchange segment AD and CB, this is translocation.

Types

Simple translocation In this type, a small segment of a chromosome is added to the end of other nonhomologous chromosomes. Often this is caused by single break in one chromosome only.

Shift translocation Here, an interstitial segment of one chromosome is broken off and is inserted within the break in another nonhomologous chromosome. Thus it involves three breaks—two in one chromosome and one in a nonhomologous chromosome.

Reciprocal translocation There is a two-way exchange of segments between the chromosomes. It is produced by a single break in each of the two nonhomologous chromosomes and exchange of parts. A reciprocal translocation between chromosomes AB ● CDEFG and MN ● OPQRS might give rise to chromosomes AB ● CDQRG and MN ● OPEFS

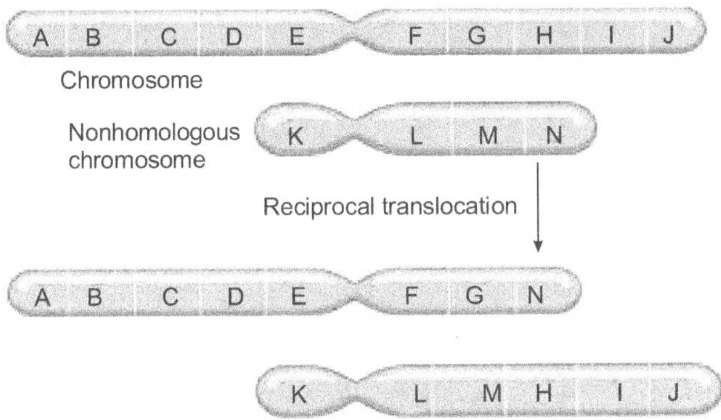

Figure 10.13   Translocation
(*Source*: www.trc.ucdavis.edu)

Multiple translocation   In this type, more than two pairs of nonhomologous chromosomes exchange parts.

Effects of translocation   During synapsis in cell division, pairing of translocated chromosomes takes up a cross-shaped configuration. Translocations result in new linkage arrangements, position effects and semi-sterility.

In a reciprocal translocation heterozygote (two normal and two translocated) chromosomes pair to form a cross-shaped configuration at pachytene stage of meiosis, whereas the normal chromosomes form a bivalent. The occurrence of crossing over and chiasmata formation in each arm of the quadrivalent, results in the formation of a ring or circle of four chromosomes at diakinesis and at metaphase of first meiotic division.

Deletions frequently accompany translocations. In a **Robertsonian translocation**, for example, the long arms of two acrocentric chromosomes become joined to a common centromere and generate metacentric chromosome. Robertsonian translocations are the cause for some cases of Down syndrome.

The anaphasic movement of chromosomes towards the poles takes place in any one of the three different ways given below.

i. *Alternate segregation*   The alternate chromosomes of the ring go to the same pole; the normal chromosomes on one pole and the translocated chromosomes on the other pole. Therefore, all the gametes receive full complement and are fully viable.

ii. *Adjacent-1 segregation*   In open ring configuration, one normal and one translocated chromosome reach one pole and one translocated and one normal chromosome the other pole. Here both types of gametes are unbalanced, carrying duplications and deletions that are usually lethal.

iii. *Adjacent-2 segregation*   In some translocations, the two chromosomes (one normal and other translocated) go to one pole and the other two go to the other pole. So, four types of gametes are formed.

Significance

i.   Translocation may cause changes in the morphology or appearance of chromosomes by centric fusion between two acrocentric chromosome segments. This may lead to a change in the number of chromosomes.

ii. Translocation introduces genetic polymorphism in the population and plays a role in the formation of new species.

Inversion

Sometimes the number of genes in a chromosome is not changed but the sequence of genes is altered by the rotation of a gene segment within a chromosome by 180° (Figure 10.14). For example, if a chromosome having gene sequence A B C D E F G H I breaks at point B and G, and the middle segment C D E F undergoes inversion; the gene sequence in the inverted chromosome will be A B F and E D C G H I J.

Types

Paracentric inversion   When both the breaks in the chromosome during inversion occur on the same side of the centromere, the inversion is known as **paracentric**. The inverted segment is without a centromere. If paracentric inversion occurs singly, i.e., on one side of the centromere alone, it is known as **intraradial** or **homobrachial** inversion. On the other hand, when two paracentric inversions occur one on either side of the centromere, the inversion is known as **interradial** or **brachial** inversion.

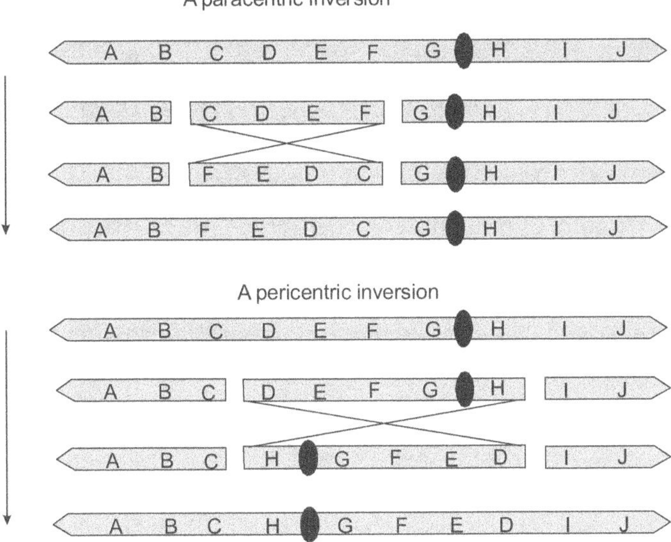

Figure 10.14   Inversion
(*Source*: www.depts.washington.edu)

Pericentric inversion   In pericentric inversion the inverted segment contains the centromere, i.e., it involves one break on either side of the centromere.

Inversions are common in many organisms like *Drosophila*, grasshoppers and mosquitoes. Inversions may have played a role in human evolution. Studies reveal that several human chromosomes differ from those of chimpanzees by only a pericentric inversion.

Effects At the time of gamete formation, inversion disrupts pairing, and abnormal configurations are obtained. The paracentric inversion with one chiasmata in the region of inversion, results in the formation of an acentric chromatid lacking centromere, and a dicentric chromatid connecting the two chromosomes. The acentric and dicentric chromatids are inviable.

Crossing over in a pericentric inversion results in the formation of the chromosomes which have duplications for certain genes and deletions for others. Pericentric inversion changes the appearance of a chromosome. A chromosome with median centromere ('V' shaped), may change into hook-shaped or rod-shaped form.

Effects of inversion on phenotype may be the occurrence of position effect. A gene in one location on a chromosome does not necessarily have the same action that it could have in another position. Inversions suppress crossing over and tend to retain the original combination of genes.

Significance Inversions help in the origin of new species. As a result of inversions, the crossing over frequency is reduced and so recombination is less.

VARIATIONS IN CHROMOSOME NUMBER

Anomalies of chromosome number occur as either euploidy or aneuploidy. Euploidy involves changes in whole sets of chromosomes; aneuploidy involves changes in chromosome number by addition or deletion of less than a whole set. Aneuploids arise by non-disjunction where there is failure of separation of chromosomes during cell division (Figure 10.15).

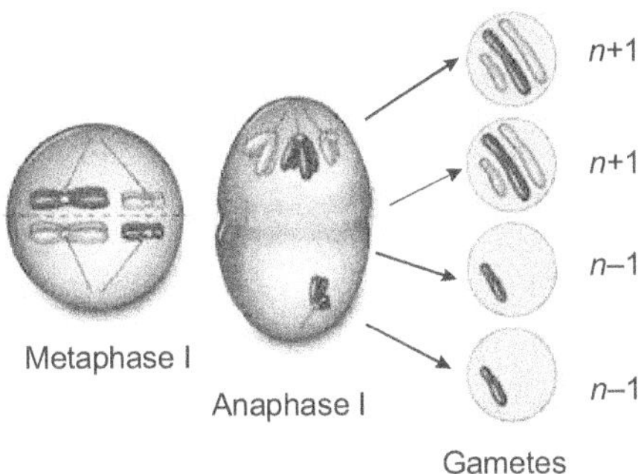

Figure 10.15 Non-disjunction
(*Source*: www.ccs.k12.in.us)

Aneuploidy

Aneuploidy may be of many kinds. A diploid cell ($2n$) missing a single chromosome ($2n - 1$) is called monosomic. A monosomic person has 45 chromosomes. A cell that lacks both copies of chromosome is termed nullisomic ($2n - 2$). In humans, for example, a nullisomic person has 44 chromosomes. A double monosomic ($2n - 1 - 1$) is a cell that lacks one chromosome from two different pairs.

With regard to the addition of a chromosome, a diploid cell with an extra chromosome is trisomic ($2n + 1$). A trisomic person has 47 chromosomes. If there are two additional chromosomes in a cell, it is tetrasomic ($2n + 2$).

Aneuploidy in humans The most common aneuploids seen in living humans are sex-chromosomal aneuploids, e.g. Turner syndrome and Klinefelter syndrome.

Autosomal aneuploids are often spontaneously aborted early in development. The common autosomal aneuploid is trisomy 21, also called Down syndrome.

Down syndrome (47, XX or XY)   It was first reported by **Langdon Down** (1866). The defect is caused by trisomy of 21st chromosome (Figure 10.16). The small extra chromosome is formed due to non-disjunctional error during the formation of gametes where the 21st chromosomes failed to separate. Most children with Down syndrome are born to normal parents. In most cases, it arises from maternal non-disjunction and the frequency of this occurring correlates with maternal age. The occurrence of trisomy in live births and in spontaneous abortions increases with the age of the mother. Researches indicate that a mother of 35 years has more risk carrying a child affected by Down syndrome. The risk is 3% higher compared to mother of 25 years.

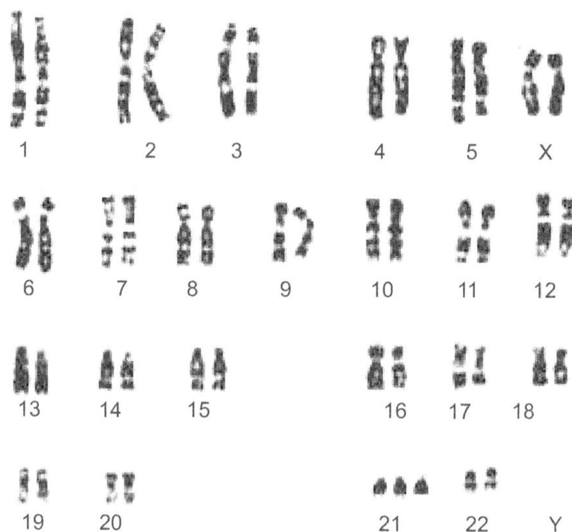

Figure 10.16   Human karyotype showing 21 trisomy

Down syndrome also occurs due to translocation of chromosomes 21 and 14 or 15. This is called **familial Down syndrome.** Since the facial features resemble that of a mongoloid race, Down syndrome is also termed as mongolism.

The features are moon face, open mouth, projecting lower-lip, long-rigid tongue, stubby hands and feet, occurrence of simian line in the palm, slanting eyes, broad forehead, congenital malformations especially of the heart, low level of calcium in blood and susceptibility to respiratory disorders (Figure 10.17). Mental retardation is observed and IQ is 25–50 compared with an average IQ of 100 in general population. Incidence is about 1/700 live births.

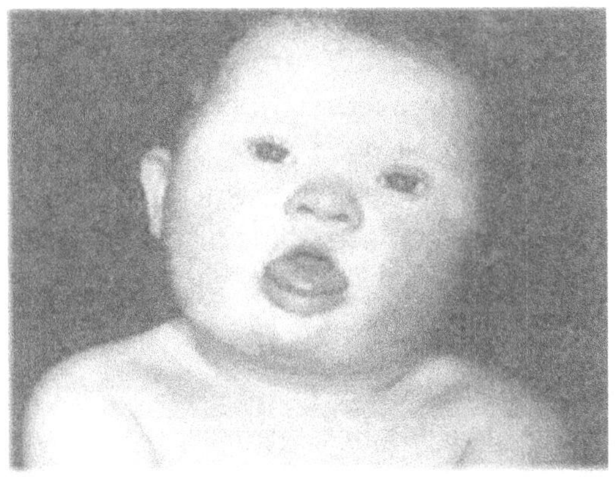

Figure 10.17 Down syndrome baby
(*Source*: www.manbir.online.com)

Trisomy 18, also known as Edward syndrome, arises with a frequency of approximately 1 in 8000 live births. Babies with Edward syndrome are severely retarded and have low-set ears, a short neck, deformed feet, clenched fingers, heart problems and other disabilities. Few live for more than a year after birth. Trisomy 13 has a frequency of about 1 in 15,000 live births and produces features that are collectively known as Patau syndrome. It has characteristics including severe mental retardation, a small head, sloping forehead, small eyes, cleft lip and palate, extra fingers and toes, and numerous other problems. About half of children with trisomy 13 die within the first month of life, and 95% die by the age of 3.

Trisomy 8 is rarer and arises with a frequency of about 1 in 25,000 to 50,000 live births. This aneuploid is characterized by mental retardation, contracted fingers and toes, low-set malformed ears, and a prominent forehead. Many who have this condition have normal life expectancy.

Klinefelter syndrome (47, XXY)   The male has one extra X chromosome, thus inhibiting the development of male characters. When an abnormal egg with XX chromosome is fertilized by a sperm carrying Y chromosome, a zygote with normal autosomes and XXY chromosomes is formed. Features are long legs, mental retardation, external genitalia are of male type but the testes are very small, body hair is sparse, feminized secondary sexual characters, female like breast development (gynecomastia) and no sperm production. They are infertile (Figure 10.18). Incidence is about 1/500 male births.

This condition can be treated by surgical removal of breast. Although sterility is unalterable, treatment with testosterone does promote development of sex organs, body hair, musculature and deeper voice.

Turner syndrome (45, XO)   The female lacks one X chromosome. This is produced when an egg with X chromosome is fertilized by a sperm without a sex chromosome. They are characterized by short stature, webbed neck, low-set ears, hypoplastic nails, under-developed breast and broad shield-like chest with widely spaced nipples, no ovary, uterus and no menstruation, hormonal abnormality, sexual hairs scanty, deformed heart, horseshoe-shaped kidneys, double ureters and mental retardation (Figure 10.18). Additional skeletal deformities are seen sometimes: a high, arched palate, receding chin, mismatching of the upper and lower teeth and abnormally low bone density. They are sterile with gonadal dysgenesis (rudimentary gonad). Incidence is about 1/5000 female births. It is highly lethal in embryos, being the most common type among spontaneous abortions and account for 20% of all chromosomally abnormal aborted embryos. 98% of Turner syndrome is lost during the first three months of pregnancy.

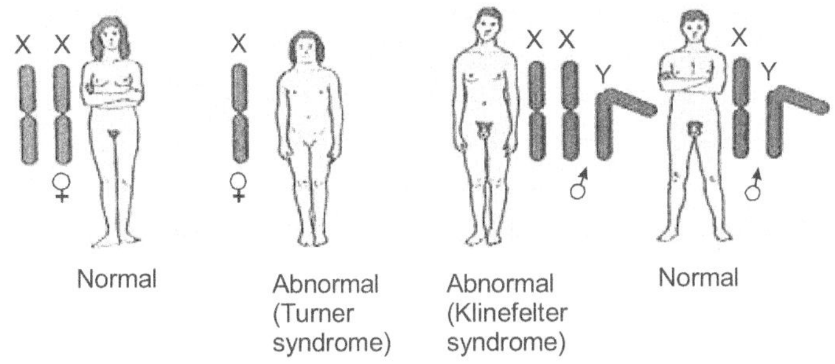

Figure 10.18 Turner and Klinefelter syndrome
(*Source*: www.emc.maricopa.edu)

Aneuploidy in Drosophila   Aneuploidy results from non-disjunction in meiosis or by chromosomal lagging whereby one chromosome moves (laggard chromosome) more slowly than others during anaphase, is excluded from the telophase nucleus and thus is lost.

Bridges first showed the occurrence of non-disjunction in Drosophila in 1916 with crosses involving the white-eye locus. When a white-eyed female was crossed with a wild type male, the daughters were wild type and the sons were white-eyed. However, occasionally, a white-eyed daughter or a wild type son appeared. This could be explained by a non-disjunctional error in the white-eyed females, where X^wX^w and O eggs (without sex chromosomes) were formed. Under this hypothesis, if a Y-bearing sperm fertilized an X^wX^w egg, the offspring would be an X^wX^wY white-eyed daughter (trisomic). If a normal X^+ bearing sperm fertilized the egg without sex chromosome, the result would be an X^+O wild type son (monosomic).

Mosaicism(C)   Rarely, an individual is made up of several cell lines, each with different chromosome numbers. These individuals are referred to as mosaics or chimeras, depending on the sources of the cell lines. Such conditions can be the result of non-disjunction or chromosomal lagging during mitosis in the zygote or in the nuclei of the early embryo. A lagging X chromosome is lost in one of the dividing somatic cells, resulting in an XX cell line and an XO cell line. In *Drosophila*, if this chromosomal lagging occurs early in development, an organism that is part male (XO) and part female (XX) develops. This may result in a fly half male and half female. A mosaic of this type with male and female phenotypes is termed **gynandromorph**.

Many sex chromosomal mosaics are known in humans, including XX/X, XY/X, XX/XY and XXX/X.

Euploidy

Euploid organisms have varying number of complete haploid chromosomal sets. Gametes are haploids (n) and somatic cells are diploids ($2n$). Organisms with higher number of sets, such as triploids ($3n$), tetraploids ($4n$) and pentaploids ($5n$) are called **polyploids**.

Polyploidy is common in plants and is a major mechanism by which new plant species have evolved. Approximately 40% of all flowering plants and 70% of grasses are polyploids. They include a number of agriculturally important plants, such as wheat, oats, cotton, potatoes, and sugar cane. Polyploidy is less common in animals, but is found in some invertebrates, fishes, salamanders, frogs and lizards.

Some groups of plants have many polyploidy members. For example, *Triticum* a genus of wheat, has members with fourteen, twenty-eight and forty-two chromosomes. Because the basic *Triticum* chromosome number is $n = 7$, these forms are diploids ($2n$), tetrapoloids ($4n$) and hexaploids ($6n$) respectively. *Chrysanthemum* with a basic number $n = 9$, has species of eighteen, thirty-six, fifty-four, seventy-two and ninety chromosomes.These species represent diploid ($2n$), tetraploid ($4n$), hexaploid ($6n$), octaploid ($8n$) and decaploid ($10n$).

There are two major types of polyploidy. They are:

1. Autopolyploidy, in which all chromosome sets are from a single species; and

2. Allopolyploidy, in which chromosome sets are from two or more species.

Autopolyploidy Autopolyploidy occurs in several different ways. For example, if a diploid gamete (abnormal) fertilizes a normal haploid gamete, the result is a triploid. Similarly, if a diploid gamete fertilizes another diploid gamete, the result is a tetraploid. A tetraploid can originate also by somatic doubling of diploid tissues. This can occur by the disruption of normal nuclear division. For example, colchicine, an alkaloid, induces somatic doubling by inhibiting microtubule formation. This prevents the formation of a spindle and thus prevents the chromosomes from moving apart during either mitosis or meiosis. The result is a cell with double chromosome number. Other factors like chemicals, temperature shock and physical shock can produce the same effect.

Since all the chromosomes are from the same species, they are homologous and their segregation produces unbalanced gametes with various numbers of chromosomes resulting in sterility. The sterility due to autopolyploidy has commercial value. The diploid bananas for example, have $2n=22$ and produce hard and inedible seeds. But triploid bananas have $3n=33$ and are sterile and so produce no seeds.

Polyploidy has been used in agriculture to produce "seedless" as well as "jumbo" varieties of crops. For example, seedless watermelon is a triploid. Its seeds are mostly sterile and do not develop. It is produced by growing seeds from the cross between a tetraploid and a diploid. Jumbo macintosh apples are tetraploid.

Allopolyploidy   Hybridization between two or more species is the cause for allopolyploidy. The following example explains how allopolyploidy can arise from two species. Suppose species 1 with chromosomes ($2n=6$) AABBCC produces gametes with chromosomes ABC and species 2 with ($2n=6$) GGHHII produces gametes with chromosomes GHI. If these gametes fuse then the hybrid formed will have 6 chromosomes ABCGHI. Since these chromosomes are different, they will not pair in meiosis and segregation is not normal and so they will be sterile. Since each chromosome has its homologue, this is also referred to as amphidiploid.

In the 1920s, **George Korpechenko** created polyploids experimentally. The cabbage *Brassica oleracea* ($2n =18$) and radish *Raphanus sativa* ($2n=18$) are important plants. Only the leaves of the cabbage and roots of the radish are normally consumed. Korpechenko produced a hybrid between them so that no part of the plant is wasted. But this hybrid was sterile. After several repeated crosses he could get a fertile plant which produced seeds. The analysis revealed that the plant was a tetraploid with ($2n=36$). The offspring are sterile. To his disappointment the new plants possessed the roots of cabbage and the leaves of radish.

Significance of polyploidy The increase in chromosome number is often associated with increase in cell size, and many polyploids are bigger in size than the normal diploids. Plant breeders have used this to produce plants with larger leaves, flowers, fruits and seeds.

Polyploidy is rare in animals. There are several reasons for this. (i) Animals have chromosomal sex-determining mechanism (ii) Most of the animals reproduce sexually and problems in meiosis will arise (iii) In animals hybridization is restricted.

Only a few human polyploids have been reported, and most died within few days of birth. Triploidy is seen in about 10% of all spontaneously aborted human feotuses (Figure 10.19).

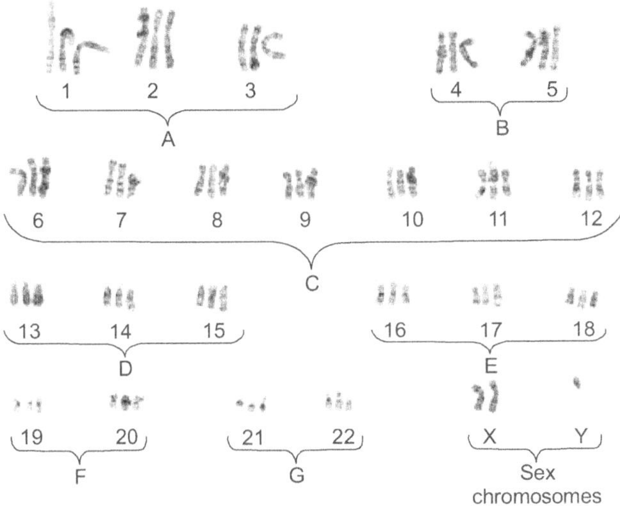

Figure 10.19 Human karyotype showing triploidy
(*Source*: www.asklenore.com)

Summary

- ✿ Mutations are heritable changes in the genetic coding instructions of DNA. Mutation rate is the frequency with which a specific mutation arises.

- ✿ Mutator genes increase the mutation frequency and the suppressor mutations are antimutators.

- ✿ Mutations may be base substitutions, insertions or deletions.

- ✿ A base substitution may be transition or transversion produced by tautomerization, ionization or base analogs. Frameshift mutations are caused by deletion or addition of nucleotides.

- ✿ Expanding trinucleotide repeats are associated with several genetic diseases.

- ✿ Depurination and deamination cause copy error mutations. Mutation-causing agents are called mutagens.

- ✿ Ionizing radiation causes DNA breaks and alters the structure of DNA. Ultraviolet rays cause dimerization and disrupt replication and transcription.

- ✿ Mutations can be detected by ClB method. Ames test is widely used to screen chemicals for mutagenicity.

- ✿ Each species of animals or plants has a fixed number of chromosomes. These chromosomes are represented one set in haploid gametes and twice in diploid body cells.

- ✿ Sometimes changes occur in the structure or number of chromosomes. The types of chromosomal variations are deletion, duplication, translocation and inversion.

- ✿ Deficiency is the deletion of a chromosomal segment resulting in the loss of genes. The presence of same block of genes more than once in a haploid complement is known as duplication. Translocation is a kind of chromosomal rearrangement in which a block of genes from one nonhomologous chromosome is transferred to another nonhomologous chromosome. Sometimes the number of genes in a chromosome is not changed but the sequence of genes is altered by the rotation of gene segment within a chromosome by 180° resulting in inversion.

✿ Anomalies of chromosome number occur as either euploidy or aneuploidy. Euploidy involves changes in whole sets of chromosomes and aneuploidy involves changes in chromosome number of addition or deletion of less than a whole set. Aneuploids arise by non-disjunction where there is failure of separation of chromosomes during cell division. The most common aneuploids seen in living humans are sex-chromosomal aneuploids. Turner syndrome and Klinefelter syndrome are the examples. The common autosomal aneuploid is trisomy 21, also called Down syndrome.

✿ Rarely, an individual is made up of several cell lines, each with different chromosome numbers. These individuals are referred to as mosaics or chimeras. Such conditions can be the result of non-disjunction or chromosomal lagging during mitosis in the zygote or in nuclei of the early embryo. A mosaic with male and female phenotypes is termed gynandromorph.

✿ Euploid organisms have varying number of complete haploid chromosomal sets. Gametes are haploids (n) and somatic cells are diploids ($2n$). Organisms with higher number of sets are called polyploids. Polyploidy is common in plants and is a major mechanism by which new plant species have evolved. There are two major types of polyploidy. They are autopolyploidy and allopolyploidy.

REVIEW QUESTIONS

1. What is the difference between transition and transversion?

2. Differentiate between the mutator gene and suppressor mutator.

3. How does frameshift mutation arise?

4. What kind of mutations can occur when cells are treated with acridine dye, alkylating agent and 5-bromouracil?

5. Propose an explanation for the occurrence of gynandromorphs in fruit flies.

6. How will you differentiate between crossing over and translocation.

7. State the implications of deletion and duplication in a chromosome.

8. Sketch the kinds of inversion.

9. What is the difference between autopolyploidy and allopolyploidy? How does each arise?

10. A species has $2n=12$ chromosomes. How many chromosomes will be found per cell in each of the following mutants in this species?

    i. monosomic

    ii. trisomic

    iii. nullisomic

    iv. tetrasomic

    v. autotriploid

    vi. autotetraploid

11

GENETIC DISORDERS

Objectives

- To envision the concept of metabolic blocks and inborn errors of metabolism
- To gain an overview of some common genetic disorders

Key Terms

phenylketonuria	alkaptonuria	tyrosinosis
goitrous cretinism	ganglioside	Tay-Sachs disease
albinism	hexaminidase	juvenile diabetes
dehydrogenase	desferal	beta-thalassemia
favism	cystic fibrosis	sickle-cell anaemia
haemolytic anaemia	sickle-cell trait	hydrops foetalis
thalassemia	alpha thalassemia	dystrophin
diabetes mellitus	intersex	gestational diabetes
mosaicism	hypertension	hermaphroditism
pseudohermaphroditism	testicular feminization	Lesch-Nyhan syndrome
homogentisic oxidase	primaquine sensitivity	G6PD deficiency
phenylalanine hydroxylase	inborn errors of metabolism	

non-insulin-dependent diabetes

thalassemia minor or Cooley's trait

thalassemia major or Cooley's anaemia

INTRODUCTION

Genes control the expression of a trait in organisms through the developmental and biochemical activities of their cells. All the cellular activities are controlled by enzymes, whose synthesis is regulated by genes. A dominant or wild type gene is capable of synthesizing the required enzyme by expressing itself, whereas the recessive gene is unable to produce the functional enzyme and hence not able to express itself. While working with *Neurospora crassa*, in 1941, **Beadle** and **Tatum** demonstrated that genes express themselves by the synthesis of enzymes. They received the Nobel Prize in 1958 for proposing the "One gene–one polypeptide" hypothesis. This hypothesis explains that there are several steps before a precursor substance is transformed and expressed as a structural or functional protein. This process constitutes the metabolic or biochemical pathways. Each step of the metabolic pathway is catalysed by a specific enzyme which in turn is synthesized under the control of a specific gene.

INBORN ERRORS OF METABOLISM

Metabolic processes are mediated by enzymes. Enzymes are proteins produced under genetic control and any change in the gene alters the products of metabolism. In 1902, **Sir Archibald Garrod** described in his book *Inborn Errors of Metabolism* that most of the hereditary diseases are caused by metabolic blocks resulting from abnormality of certain specific enzymes and responsible for many of the genetic disorders.

Metabolic Errors in Phenylalanine Metabolism

Phenylketonuria (PKU) Five metabolic diseases have been noticed in human beings associated with phenylalanine, an essential amino acid of dietary proteins. PKU is caused by a rare recessive autosomal mutant gene designated as pp.

Biochemical block is formed due to the absence of the enzyme, phenylalanine hydroxylase, which catalyses the conversion of phenylpyruvic acid into hydroxy phenylpyruvic acid. As a result, phenylpyruvic acid accumulates in the blood and gets deposited in the vital organs of the body and causes impairment of brain tissue. PKU is characterized by light hair, skin and eye colour, mental retardation and microcephaly. Folling first reported PKU and so is also called as Folling's disease. It affects 1 in 11,000 newborns. PKU was found to account for about 1% mentally retarded patients in institutions. PKU is also linked to other metabolic blocks. Figure 11.1 depicts the five inborn errors of metabolism

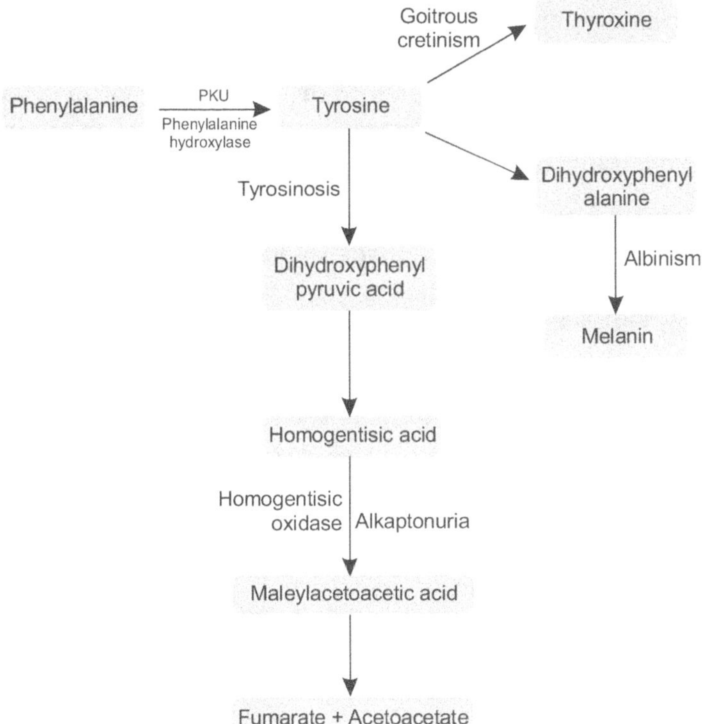

Figure 11.1   Metabolic errors in phenylalanine metabolism

resulting from enzyme blocks at five points in the breakdown of phenylalanine—a building block of protein. PKU results from a block in the first step of pathway.

PKU can be treated by diet therapy. PKU patients are given low phenylalanine diet comprising of apple, banana, cabbage, carrots, tapioca, corn, sugar, butter, and multi-vitamin supplements.

Alkaptonuria Persons suffering from alkaptonuria excrete urine that turns black on exposure to air. This is due to the presence of a substance alcapton homogentisic acid in their urine. The defect in them is the absence of the enzyme homogentisic oxidase due to the autosomal recessive mutation hh. This enzyme deficiency results in the failure of conversion of homogentisic acid to maleylacetoacetic acid. The homogentisic acid is passed into the urine which turns black when exposed to air. Alkaptonuria causes severe joint pain, black deposits on palate, ears and eyes.

Tyrosinosis The recessive gene tt blocks the conversion of hydropyruvic acid into dihydroxyphenyl pyruvic acid. This leads to the accumulation of tyrosine which is excreted in the urine. This is harmless.

Goitrous cretinism In this condition due to the mutation cc the enzyme that converts tyrosine into thyroxine is absent and the metabolic pathway is blocked. Tyrosine accumulates in the urine. The disease is characterized by mental retardation, stunted growth and hypertrophy of thyroid gland.

Albinism A block in the next step leads to accumulation of the amino acid tyrosine and lack of its product, the pigment melanin, causing pink eyes, white hair and white skin of albinism. The persons with genotype aa fail to produce the enzyme required for the conversion of dihydroxy phenylalanine (DOPA) into melanin and such persons are albinos.

Tay-Sachs Disease (TSD)

This disease is a metabolic disorder of lipid metabolism. It is characterized by a "cherry red spot" in the eye. When the child is 1 year old, it suffers from generalized paralysis, blindness, gradual loss of hearing, severe feeding difficulties and some enlargement of the head. By 2 years, the child is completely immobile and most of them die at the age of 4.

The nerve cells of the brain become enlarged due to the accumulation of the lipid ganglioside called G_{M2}. The basic defect is a missing enzyme hexaminidase which normally splits off the sugar group from the rest of the ganglioside molecule. TSD results from the autosomal recessive mutation of the gene *HEX-A*.

Lesch-Nyhan Syndrome (LNS)

This is caused by error in the nucleic acid metabolism. In this disease the purines are not recycled by salvage pathway, instead excess purines are converted to uric acid. The enzyme hypoxanthine-guanine phosphoribosyl transferase (HGPRT) which is necessary for the conversion, is absent due to a sex-linked recessive mutation. The child with LNS suffers from various problems: extremely high concentrations of uric acid in the urine and blood, mental retardation, constant and involuntary writhing movements of the legs and arms and self-mutilation (causing damage to their own body parts). Early in childhood, the occurrence of orange sand (uric acid crystals) in the urine is noted. Treatment for LNS includes high fluid intake, adequate nutrition and use of drug.

G6PD Deficiency

Glucose 6-phosphate dehydrogenase is an important enzyme of the red blood cells and its deficiency is inherited as an X-linked recessive trait. The gene responsible for the G6PD deficiency is located on the X chromosome.

This enzyme is necessary for the metabolism of carbohydrates. Deficiency of the enzyme results in sensitivity to sulpha drugs, anti-malarial drugs—primaquine (primaquine sensitivity), antipyretics, naphthalene, fava beans, etc. and produces haemolytic anaemia. Favism is a haemolytic condition produced in a person with G6PD deficiency by eating fava beans.

It is estimated that more than 150 million people suffer from G6PD deficiency all over the world. In the Indian population groups, the association between G6PD deficiency and neonatal jaundice has been well established. The distribution of the deficiency has been shown to be connected with the prevalence of falciparum malaria. It has been suggested that G6PD deficiency confers some resistance to falciparum malaria.

DISORDERS ASSOCIATED WITH HAEMOGLOBIN

Sickle-Cell Anaemia (HbSS)

Sickle-cell anaemia is a condition in which the affected individual possesses an abnormal variant of adult haemoglobin. The red blood cells are sickle-shaped. This is caused by a gene mutation which alters the quality of the globin. Haemoglobin has four polypeptide chains in its protein part. Adult haemoglobin is made up of two alpha and two beta globin chains. Foetal haemoglobin is composed of two alpha and two gamma globin chains. In sickle-cell anaemia, one of the amino acids, is altered in the beta globin of the haemoglobin. In normal haemoglobin the 6th amino acid is glutamic acid while in sickle cell it is substituted by valine by changing the codon GAG to GTG. Persons with sickle-cell anaemia have an abnormally large DNA fragment when the beta globin gene is cut with appropriate restriction enzyme (Figure 11.2).

This disease resulting from homozygous recessive condition, is a haemolytic condition (shortened lifespan of red blood cell) leading to severe anaemia. This disease characterized by enlarged spleen, painful cries, organ damage, impaired mental functions, increased susceptibility to infection and early death. The sickle-cells block the vessels and disturb normal circulation in blood vessels.

The wild type is HbAA. The heterozygous condition (HbAS) is termed as sickle-cell trait. The inheritance of sickle-cell disease obeys the principles of Mendelian inheritance. The cross HbAS × HbAA has equal chance of producing offspring with either normal (HbAA) or sickle-cell trait (HbAS). If both parents have sickle cell trait (HbAS × HbAS) then the offspring may have 50% chance of sickle-cell trait, 25% chance of sickle-cell anaemia (HbSS) and 25% chance of normal (HbAA).

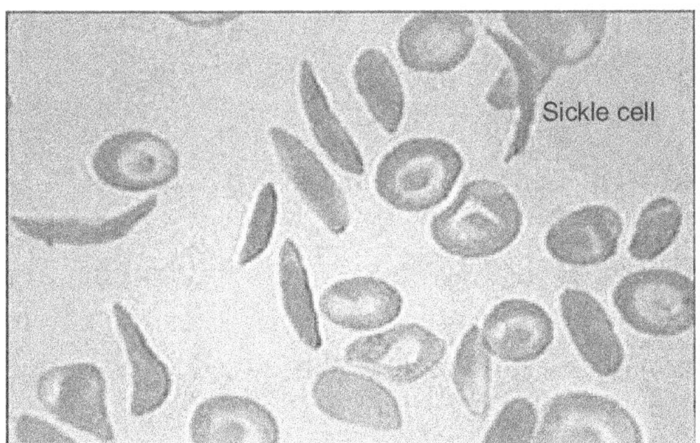

Figure 11.2   Normal red blood cells and sickle cells
(*Source*: www.revolutionhealth.com)

Very high frequency of sickling has been reported from various parts of south, western and central India. It has been estimated that over 5,00,000 persons among tribals alone are carriers of this genetic disorder and over 2,00,000 are homozygous.

The following steps should be taken for the prevention and treatment of sickling crisis.

The patient should be kept warm, avoiding drinking cold water. To prevent dehydration, adequate fluid must be taken and excessive heat is avoided. Clot formation within the blood vessels may be prevented by using drugs. Blood transfusion is given when the haemoglobin level falls. Prophylactic antibiotics are given to prevent infection.

Thalassemia

Thalassemias are haemoglobinopathies and represent a group of disorders affecting haemoglobin synthesis, i.e., the production of normal haemoglobin is inhibited. Thalassemia is a common disorder affecting many people in India. The two commonly found thalassemias are beta thalassemia in which beta globin chain synthesis is reduced and alpha thalassemia where alpha chain synthesis is affected. Both can exist in homozygous or heterozygous states.

Beta thalassemia in homozygous state is known as **thalassemia major** or **Cooley's anaemia**. In this disorder, red cells of varying size with distorted forms termed as **poikilocytes** occur. Severe anaemia occurs requiring frequent blood transfusions and is often fatal in early childhood. RBC count is much reduced. Splenomegaly is found.

Beta thalassemia in heterozygous state is known as **thalassemia minor** or **Cooley's trait**. These people are perfectly healthy.

Alpha thalassemia is a rare disease. In alpha thalassemia major, both alpha chain production is affected and the condition is hydrops foetalis which is fatal. Alpha thalassemia minor (heterozygous state) is harmless.

Thalassemia in some form or other is one of the important haemoglobinopathies in India. The disease has been reported from almost all regions and is clearly most widespread. Several studies have brought to light a large number of subjects with thalassemia in various forms.

The only treatment for thalassemia is regular blood transfusions, usually every four weeks and bone marrow transplant. The drug Desferal is given to remove excess of iron when blood transfusion is done frequently.

OTHER GENETIC DISORDERS

Duchenne Muscular Dystrophy (DMD)

DMD is an X-linked recessive disease occurring in about 1 in 3,500 newborn males. The normal muscle protein dystrophin is absent. Usually the onset occurs before the age of 6 years, and the victim is chair-ridden by age 12 and dead by age 20. The earliest signs are delays in walking and talking. Then come weakness and gradual wasting of thigh and pelvic muscles, leading to unsteadiness, difficulty in walking stairs or rising from chairs. As the shoulder, trunk and back muscles gradually weaken and degenerate, the child develops a swayback posture, has trouble maintaining balance and falls a lot. Heart problems are the rule, sometimes leading to sudden death. DMD is often caused by a break in the short arm of X chromosome, i.e., Xp 21.

Cystic Fibrosis

It is a lethal autosomal recessive disorder. About 1 in 2500 newborns in Europe is affected by cystic fibrosis. About 1 in 25 is heterozygous and is a carrier for this disease. There are serious respiratory and digestive problems and extremely salty sweat with high sodium and chloride levels ("salty kisses" in children). Thick, sticky mucus secretion clogs tubules of lungs and other organs causing irreversible damage and malfunctioning of lungs, liver, pancreas, intestine, sweat glands and reproductive organs. Growth is retarded in children. Because of lung problem they are highly susceptible to infection. The gene responsible for this disease is CF and is located in the long arm of chromosome 7. It encodes a protein termed as cystic fibrosis transmembrane conductance regulator (CFTR) that regulates the movement of chloride ions across the cell membrane. Patients with cystic fibrosis have a mutated, dysfunctional CFTR. Average life expectancy for a person with CF is 20 years.

The best treatment includes regular chest physiotherapy to loosen secretions, antibiotics and replacement of digestive enzymes.

Diabetes Mellitus

Diabetes mellitus is a condition characterized by faulty sugar metabolism due to lack of the hormone insulin secreted by the beta cells of the pancreas. It is characterized by increased sugar level and weakness. Frequent urination is often present. The age of onset may vary from childhood to very old age. The expression of the disease may also vary.

There are two major types of diabetes.

Type I   Insulin-dependent and often called juvenile diabetes.

Type II Non-insulin-dependent, and often called adult onset.

There is also gestational diabetes occurring only in pregnancy. In type I, the pancreas does not make enough insulin. Without taking insulin the affected person will die. Type II diabetes is 10 times more prevalent than type I. It is one of the most common chronic diseases in India. It is associated with obesity and aging. It is a lifestyle disease.

Here insulin is secreted but the cells are unable to utilize it for metabolizing sugar. Incidence is high in the age group of 40 to 70 years. Above the age of 50 years it is more in females than in males.

Three theories have been put forward regarding its inheritance.

   i. Autosomal dominant inheritance

  ii. Autosomal recessive inheritance

 iii. Multifactorial inheritance

Studies have shown that diabetes is not simple; it is genetically complex, involving multiple genes, and multiple gene–environment interactions. The Human Genome Project has identified chromosome 2 carrying a gene locus showing susceptibility to type II diabetes. Two new susceptibility genes have been localized in 11q and 6q.

Hypertension

Hypertension is the measure of pressure or tension exerted by the blood on the walls of arteries, and elevated level of pressure is called hypertension or high blood pressure. In a normal adult, blood pressure rises to a peak value (systolic pressure) 120 mmHg in each cardiac cycle and falls to a minimum value (diastolic pressure) of about 70 mmHg. High blood pressure is 140/90 mmHg. Hypertension affects 25% of the world's adult population and is a major risk factor for stroke, myocardial infarction and heart and kidney failures. Although scientists have believed this condition to be hereditary, it is the result of a combination of hereditary and environmental factors. The Human Genome Project has identified genes in chromosomes 2, 5, 6 and 15 responsible for high blood pressure.

Intersexes

Intersex condition is one in which the genitalia are ambiguous and the sexual characteristics of a person may not be according to his or her chromosome pattern. There are two ways of expression of intersex. One is true hermaphroditism and the other is pseudohermaphroditism.

A true hermaphrodite is an individual who has the reproductive organs of both sexes. The external genital differentiation is usually incomplete and anomalous. The gonad may be ovary or testis separately or combined in an ovo-testis. Externally they are more male. 80% of the patients develop gynecomastia and about 50% of the patients menstruate. There is mosaicism with regard to chromosomes. The individuals have more than one kind of chromosome complement. For example, 46XX/47XXX; 45XO/46XX; 46XX/48XXYY, etc.

There is "testicular feminization" when pseudo-hermaphroditism is seen in a male. These patients are phenotypically and physiologically Y males who are amenorrhoeic and infertile. There is no sperm production. The chromosome complement is normal. If it occurs in a female the development of internal genital organs are normal while the external genitalia show varying degrees of virilization. There is primary amenorrhoea.

CONSANGUINEOUS MARRIAGES AND GENETIC DISEASES

It is well established that the children of consanguineous marriages (marriages between related individuals) are much more likely to suffer from malformations and genetic diseases. The nature of consanguinity consists of the fact that relatives, because they have common ancestors, possess a greater number of common genes. The number of common genes depends upon the degree of relationship. For example, the share of common genes due to common ancestry in the case of

i. identical twins (monozygotic twins) $=1$;

ii. parent–children $=1/2$;

iii. uncle–niece/aunt–nephew $= 1/4$;

iv. first cousins $= 1/8$;

v. second cousins $= 1/32$.

There is increased risk of genetic disorders in children born out of consanguineous marriages. There is reduction in heterozygosity of deleterious genes and increase in homozygosity among the children of related (first cousins) individuals. There is higher risk with regard to the occurrence of early foetal death, congenital malformations, mental defects, deaf-mutism, albinism, phenylketonuria, xeroderma pigmentosum, total colour blindness, etc. in consanguineous marriages.

Summary

- Genetic disorders are caused by gene mutations. Metabolic processes are mediated by enzymes.

- Enzymes are proteins produced under genetic control and any change in the gene, alter the products of metabolism.

- Inborn errors of metabolism are caused by metabolic blocks.

- Five metabolic diseases have been noticed in human beings associated with phenylalanine, an essential amino acid of dietary proteins.

- PKU is caused by a rare recessive autosomal mutant gene resulting in the absence of the enzyme phenylalanine hydroxylase.

- Alkaptonuria is the presence of a substance alcapton in the urine. The defect is due the absence of the enzyme homogentisic oxidase.

- Tyrosinosis is caused by the accumulation of tyrosine.

- Goitrous cretinism is due to the metabolic block in the conversion of tyrosine into thyroxine.
- Lack of the pigment melanin causes albinism.
- Tay-Sachs disease is a metabolic disorder of lipid metabolism caused by the absence of the enzyme hexaminidase.
- Lesch-Nyhan syndrome is caused by an error in the nucleic acid metabolism resulting in the absence of the enzyme HGPRT.
- G6PD deficiency is inherited as an X-linked recessive trait and the gene responsible for the G6PD deficiency is located on the X-chromosome, and produces haemolytic anaemia.
- Cystic fibrosis is a lethal autosomal recessive disorder, where the affected person has a mutated, dysfunctional CFTR.
- Sickle-cell anaemia is a condition in which the red blood cells are sickle-shaped caused by amino acid substitution.
- Thalassemias are haemoglobinopathies affecting haemoglobin synthesis.
- Duchenne muscular dystrophy is an X-linked recessive disease caused by a break in the short arm of X chromosome.
- Cystic fibrosis is a lethal autosomal recessive disorder.
- Diabetes mellitus is of 2 types—Type I, is insulin-dependent or juvenile diabetes. Type II, is non-insulin-dependent or adult onset.
- Hypertension is the result of combination of hereditary and environmental factors.
- Intersex condition is one in which the genitalia are ambiguous and the sexual characteristics of the person may not be according to his or her chromosome pattern.
- It is well established that the children of consanguineous marriages suffer from malformations and genetic diseases. There is reduction in heterozygosity of deleterious genes and increase in homozygosity.

REVIEW QUESTIONS

1. Show the biochemical blocks observed in the degradation of phenylalanine.

2. Discuss the "one gene–one polypeptide" concept using an example.

3. State the genetic cause for the following diseases.

    i.   Tay-Sachs disease

    ii. Lesch-Nyhan syndrome

    iii. Cystic fibrosis

    iv. Duchenne muscular dystrophy

4. Explain the genetic disorders related to haemoglobin production.

5. What is primaquine sensitivity and favism?

6. How is intersex expressed?

7. Mention the genetic effect of consanguinity.

12

CANCER GENETICS

Objectives

- ✿ To understand the heritable nature of cancer
- ✿ To analyse the genes involved in cancer
- ✿ To know the cancer-causing agents

Key Terms

immortalization	transformation	metastasis
malignant	benign	leukaemias
sarcomas	carcinomas	*BRCA*1
BRCA 2	mutagen	carcinogen
clonal evolution	oncogenes	RNA tumour viruses
proto-oncogenes	anti-oncogene	*p*53 gene
oncogenic viruses	DNA tumour viruses	co-carcinogen
papilloma virus	adenoviruses	herpesviruses
polyoma virus	biotransformation	
Xeroderma pigmentosum	retinoblastoma gene	
tumour-suppressor genes		

INTRODUCTION

Cancer is the second most frequent cause of death after heart disease. Cancer is not a single disease. It is a heterogeneous group of disorders characterized by the presence of cells that do not respond to the normal controls on cell division.

The cancer cells divide continuously and rapidly and produce tumours. In an advanced state, the cells of the tumour can travel to distant regions in the body and develop new tumours. They are characterized by immortalization (indefinite growth), transformation and metastasis (migration and invasion of normal tissues). When a tumour is able to invade other tissues, it is termed as malignant which is different from benign, the tumour that is not able to invade other tissues. Benign tumours are non-cancerous growth, and can be removed surgically. Malignant tumours may spread throughout the body threatening the patient's life. Certain people may have a genetic predisposition to cancer. Many environmental factors such as ionizing radiation, exposure to mutagenic chemicals and viruses, and nutritional excesses or deficiencies contribute to increased risk of cancer formation.

Cancer can be defined as the "unregulated growth and production of cells". Normal cells grow, divide and die and are regulated by internal and external signals. But in cancer cells some of these signals are disrupted and so the cells divide uncontrollably. The cancer cells lose their normal shape and become a mass of abnormal cells, which is termed as tumour. Often these cells are incompletely differentiated. Cancers can be subdivided into leukemias and lymphomas, cancer of white blood cells; sarcomas, cancer of bone and muscle cells; and carcinomas, cancer of skin.

CANCER AND HEREDITY

Most of the human cancers are not hereditary and not contagious. But cancer cells have been genetically altered in one or more ways. There are abundant experimental evidences to support the idea that somatic mutations can cause the conversion of a normal cell into a cancer cell. Somatic-cell mutations are not passed on to the progeny; so it is not possible that all cancers can be inherited, e.g. breast cancer. Frequent cases of breast cancer are associated with mutations of breast cancer gene, *BRCA1*. This gene is located in q21 of the

17th chromosome. This gene is a large one spanning about 100,000 bases. Its protein product is of 1863 amino acids. Researches have discovered more than 100 different mutations of *BRCA*1 spread out along the gene. The incidence of breast cancer is 1/500 females. Another gene *BRCA*2 on chromosome 13 was discovered in 1992 causing a breast cancer risk in men.

It was recognized in early observations that cancer might result from genetic damage. Agents such as ionizing radiation and chemicals that cause mutation (mutagen) can produce cancer too (carcinogen). The idea that cancer begins with genetic changes is supported by the study on a rare human skin cancer, Xeroderma pigmentosum. The skin cells of these persons contain a mutant gene and consequently lack the enzymes involved in repair of DNA that has been damaged by UV light. People with this hereditary disease when exposed even to slight amounts of sunlight develop skin tumours. Some cancers are associated with particular chromosome abnormalities. For example, about 90% of people with chronic myeloid leukaemia have a reciprocal translocation between chromosome 22 and chromosome 9. Some specific types of cancers run in families. The rare childhood cancer of the eye, retinoblastoma is inherited as an autosomal dominant trait in some families suggesting that a single gene may be responsible for the cancer.

Alfred Knudson (1971) proposed a model to explain the genetic basis of cancer. He studied retinoblastoma, the eye cancer, that usually develops in only one eye. But rarely it appears in both eyes and appears at early age, and affected children often have relatives with retinoblastoma. He suggested that retinoblastoma is caused by two genetic defects, both of which are necessary for the cancer to develop. He explained that when retinoblastoma occurs in one eye, a single cell in one eye undergoes two successive mutations. Because the chance of these two mutations occurring in a single cell is remote, retinoblastoma is rare and develops in only one eye. In the case of retinoblastoma in both eyes, he proposed that the child inherited one of the two mutations required for the cancer and so every cell contains this initial mutation. For developing cancer, one cell in the eye should undergo a second mutation. Since each eye possesses millions of cells, it is possible that the second mutation could occur in at least one cell of each eye, producing cancer in each eye at a very early age.

Cancer is the result of a multistep process that requires several mutations. If one or two required mutations are inherited, then fewer additional mutations are needed to produce cancer and the cancer will run in the family. The genetic theory of cancer by Knudson has been confirmed in most cancers.

CLONAL EVOLUTION OF CANCER

Cancer arises when a single cell undergoes mutation and thereby it multiplies at a faster rate. This proliferation of cells gives rise to a clone of cells carrying the same mutation. New mutations that arise in the clone will enhance the proliferation. In this process called clonal evolution, the tumour cells that acquire more mutations will divide abnormally (Figure 12.1). The rate of clonal evolution depends on the frequency with which new mutations arise. In most cancers, the genes that repair DNA are often mutated. DNA repair mechanisms usually eliminate many of the mutations and without DNA repair, the mutations persist in all genes including those involved in cell division. For example, Xeroderma pigmentosum is caused by a defect in DNA repair, and skin cancer develops due to mutation when these people are exposed to sunlight.

Gene mutations that affect chromosomal segregation are also factors contributing to clonal evolution of tumours. Aneuploidy is noted in many cancer cells. It is evident that duplication of some genes and elimination of others may contribute to cancer.

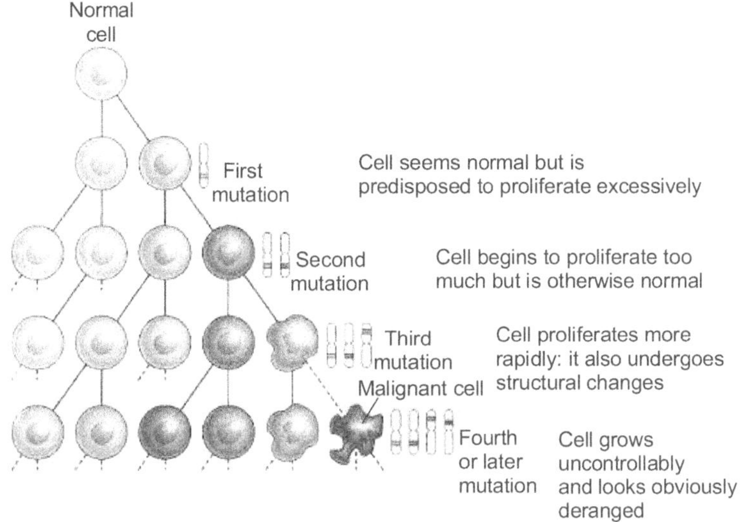

Figure 12.1   Clonal evolution of tumour
(*Source*: www.chemcases.com)

Events in the Origin of Colorectal Cancer

Colorectal cancer is an excellent example to study how cancer arises by successive accumulation of genetic defects. Some of the genes responsible for its clonal evolution have been identified. About 75% of colorectal cancers have mutations in tumour-suppressor gene *p53*, and many also have a mutation in the *ras* proto-oncogene. A defect in a gene *APC* is also noticed. Other genes that frequently mutate in colorectal cancers are the oncogenes *myc* and *neu* and the tumor-suppressor gene *HNPCC*. Mutations in these genes are responsible for the stepwise development of colorectal cancer. The first step is the inactivation of *APC* gene which increases the rate of cell division resulting in the formation of polyps. Mutations in *ras* oncogene occur later, in larger polyps comprising cells that have acquired some genetic mutations.

In normal cells the *ras* proto-oncogene sends signal from growth factors that stimulate cell division. When *ras* gene is mutated, the protein that it encodes continuously sends signals for cell division even in the absence of growth factors. Later in tumour progression, mutation in *p*53 and other genes appear. These mutations allow a cell to acquire further gene and chromosome mutations which then contribute to further proliferation and invasion into surrounding tissues. The above-explained steps are a common pathway by which colon and rectal cells become cancerous.

GENES INVOLVED IN CANCER

Cell division is regulated by two types of signals: molecules that stimulate cell division and those that inhibit it. Cancer can arise from mutations of either type of signals. A stimulatory gene may become hyperactive and such genes are always dominant. These genes are called oncogenes (Table 12.1). Cell division may also be activated when inhibitory genes become inactive. These inhibitory genes are recessive and are termed as tumour-suppressor genes.

Table 12.1   Oncogenes and human cancers

Oncogene	Cancer
myc	Burkitt lymphoma
myc	T-cell leukaemia
bcl-1	Chronic B-cell leukaemia
bcl-2	Follicular lymphoma
abl	Mylogenous leukaemia
abl	Lymphocytic leukaemia

Oncogenes and Tumour-Suppressor Genes

The first oncogene called *src* was isolated from the Rous sarcoma virus in 1970. It was **Peyton Rous** who first identified a virus that caused sarcomas in chickens and is known as Rous sarcoma virus. The virus carried a cancer-causing gene *src* that was transferred to host cell. More than 70 oncogenes have been discovered. **Michel Bishop**, **Harold Varmus** and their colleagues (1975) developed probes for several viral oncogenes and searched for related sequences in normal cells. They discovered some DNA sequences in normal cells that closely resembled the oncogenes and named them as proto-oncogenes. These genes are responsible for normal cellular functions but when they mutate they cause cancer as they act like oncogenes. In a virus infection, a proto-oncogene in a cell incorporates into viral genome by recombination. In the genome of the virus, the proto-oncogene may mutate and when inserted back into the cell, cause rapid cell division and cancer.

The conversion of proto-oncogene into oncogene in a viral genome happens in many ways. When the proto-oncogene is getting incorporated into the viral genome, its sequence may be altered or truncated. This mutated sequence may then produce abnormal protein that triggers uncontrolled cell division. The other possibility is the over-expression of gene. This can occur when the proto-oncogene by chance lies next to an enhancer during recombination. Lastly, when a virus is inserting its own DNA into the gene, the function of the proto-oncogene may be altered and the result is the disruption of normal function.

Tumour-suppressor genes are normal recessive genes that function to prevent the development of tumours. They are also termed as **anti-oncogene**. Tumours develop in cells in which both normal alleles of a tumour-suppressor gene have been lost or inactivated. The first identified tumour-suppressor gene was the retinoblastoma gene (*Rb*). In 1985, **Raymond White** and **Webster Cavenne** showed that large segments of chromosome 13 were missing in cells of retinoblastoma tumours, and isolated tumour-suppressor genes from them. The *Rb* gene is a good example of a tumour-suppressor gene. Normal activity of *Rb* gene prevents the development of retinoblastoma, which is a malignant tumour of the retina, develops in early childhood and is usually fatal if not treated. Retinoblastoma develops when both alleles of Rb tumour-suppressor gene are lost or inactivated by mutation.

The *p*53 gene is a tumour-suppressor gene that functions in normal cells. The *p*53 gene produces a DNA-binding protein. In colon cancer, increased amounts of *p*53 protein occur indicating that the action of this gene promotes tumour development. Mutation or loss of the *p*53 gene is the most common genetic change that occurs in wide variety of cancers including those of colon, breast, lung and brain.

Genes Controlling the Cell Cycle

The cell cycle is regulated by cyclins, whose concentration often fluctuate and cyclin-dependent kinase (CDK) whose concentration on the other hand remains constant. Cyclins bind to CDKs, producing activated protein kinases that initiate important events in the cell cycle. The genes that encode cyclins, and factors that stimulate or inhibit the formation of activated CDKs are often oncogenes and tumour-suppressor genes, respectively. The cancers of immune system, breast, stomach and oesophagus often have mutated genes for cyclins. The genes such as *p*16 and *p*21 are the genes that encode inhibitor factors of CDKs and they are mutated or missing in many of the cancer cells.

There are some proto-oncogenes and suppressor genes playing a role in apoptosis. Cancer cells often have mutations, DNA damage, and other cellular abnormalities that would normally stimulate apoptosis. These cells often have mutations in genes that regulate apoptosis, and so the cells do not die by apoptosis. The $p53$ gene is an example for genes that initiate apoptosis when there is DNA damage in cells. In many cancers the $p53$ gene is inactivated.

DNA Repair Genes

Defects in genes that encode components of DNA repair system have been associated with a number of cancers, e.g. Xeroderma pigmentosum. About 13% of colorectal, endometrial and stomach cancers have cells that are defective in mismatch repair system in cells. The normal allele provides sufficient amount of protein for mismatch repair. But when this allele mutates or is lost in few cells, there is no mismatch repair. These cells undergo higher-than-normal rates of mutation which leads to defects in oncogenes and tumour-suppressor genes and thus causes abnormal cell multiplication.

Genes Affecting Chromosome Segregation

A variety of chromosomal anomalies like addition of chromosomes, loss of chromosomes and chromosomal rearrangements are common features in most of the tumour cells. Aneuploidy arises when chromosomes do not segregate properly in mitosis. The proper assembly of the mitotic spindle is monitored by check points in the cell cycle. The anaphase stage is blocked when chromosomes are not properly attached to microtubules in metaphase. Some cancer cells contain mutant genes that encode for proteins involved in check point. In these cells there is improper assembly or lack of assembly of the spindle, and chromosome abnormalities result.

The tumour-suppressor gene *p53*, in addition to controlling apoptosis, plays a role in the duplication of the centrosome, which is required for the proper formation of the spindle and for chromosome segregation. Normally, the centrosome duplicates once per cycle. If *p53* is mutated or missing, the centrosome undergoes many duplications causing unequal segregation of chromosomes.

Genes that Regulate Telomerase

Telomeres are special sequences at the ends of eukaryotic chromosomes. In somatic cells, DNA replication carried out by DNA polymerase requires 3′-OH group to add new nucleotides. Because of this, the ends of chromosomes cannot be replicated, and telomeres become shorter with each cell division. This shortening finally leads to the destruction of the chromosome and cell death. So the somatic cells are capable of limited number of cell divisions. In germ cells on the other hand, the enzyme telomerase replicates the chromosome ends and maintains the telomeres. This enzyme is inactive in somatic cells. In many tumour cells, the telomerase gene is mutated, so the enzyme is expressed, and the cell is capable of unlimited cell division.

Genes that Spread Tumours

Nutrients and oxygen are essential for survival and growth of cancer cells. This is carried out by the growth of new blood vessels (angiogenesis) which is important for tumour progression. Angiogenesis is regulated by proteins and growth factors encoded by genes that are precisely regulated in normal cells. In tumour cells these genes are over-expressed, and inhibitors of angiogenesis-promoting factors may be inactivated or under-expressed.

Often primary tumours spread to distant sites and produce secondary tumours. This process of metastasis is the cause of death in 90% of human cancer cases. This is influenced by cellular changes induced by somatic mutation. Researchers have identified using microarrays, several genes that are transcribed at higher rates in metastatic cells.

CANCER-CAUSING AGENTS

The term carcinogen refers to any substance that is an agent directly involved in the promotion of cancer. Carcinogens may increase the risk of getting cancer by altering cellular metabolism or damaging DNA directly in cells, which interferes with biological processes and induces the uncontrolled, malignant division, ultimately leading to the formation of tumours. Many natural carcinogens and a large number of synthetic chemicals have been identified as carcinogens. Co-carcinogens are chemicals that do not necessarily cause cancer on their own, but promote the activity of other carcinogens in causing cancer. After a carcinogen enters the body, the body makes an attempt to eliminate it through a process called biotransformation. The purpose of these reactions is to make the carcinogen more water-soluble so that it can be removed from the body.

Radiation

Higher incidence of leukaemia among the survivors of Hiroshima and Nagasaki atomic bomb blasts indicates that atomic radiation is a carcinogen—an agent that causes cancer. UV rays damage DNA by the covalent linking of thymine bases to form dimers and a mutation arises there. Ionizing radiation like X-rays possess energy to knock electrons out of their orbits around atoms and produce mutations in the DNA of cells.

Chemicals

Many chemicals penetrate cells and interact with DNA, thereby increasing the mutation frequency. Our modern lifestyle is intimately linked to the use of chemicals. Every day almost everybody uses paints, plastics, pharmaceuticals, pesticides, etc. We use styrofoam cups, polyurethane foam cushions, plastic combs, pens and wear nylon, rayon and polyester clothes. All of us are exposed to varying amounts of thousands of chemicals used in industry and agriculture. Though most of the chemicals that we use are safe, some are proven carcinogens.

Evidence that certain chemicals cause cancer is reported from the occurrence of cancer among industrial workers. It was observed that there is a high incidence of bladder cancer in rubber industry workers, lung cancer among asbestos and arsenic industry workers and liver cancer in the workers of PVC industry. Deaths from lung cancer are almost directly proportional to the number of cigarettes smoked.

Oncogenic Viruses

The genetic information in animal viruses is carried by either DNA or RNA. These viruses can cause malignant tumours in animals and are referred to as oncogenic viruses. When an animal cell is infected by an oncogenic RNA virus, conversion of viral RNA into viral DNA and the integration of the viral DNA into the cell's chromosome are essential for continued RNA virus reproduction. Once the viral DNA has been integrated into a chromosome, it is called provirus.

The oncogene hypothesis proposes that the viral genes are passed on, along with normal genes, generation after generation. If mutagenic agents damage the DNA and trigger the synthesis of new viruses, cells containing those viruses may begin to grow in an unregulated manner and eventually develop into a tumour.

It is now known that the *src* gene carried by the Rous sarcoma virus is the gene responsible for changing a normal cell into a cancer cell. Many other oncogenic RNA viruses have been isolated from tumours in various animals. It has been discovered that both animal and human cells contain genes whose base sequences closely resemble the base sequences in the viral oncogenes and are called proto-oncogenes.

There are two classes of tumour viruses identified based on their genome. They are the DNA tumour viruses and RNA tumour viruses and the latter is also termed as retroviruses. Some of the tumour viruses are given below.

Papilloma virus   This virus causes 16% of female cancers worldwide and 10% of all cancers. They are wart-causing viruses.

Polyoma virus   SV40 is a monkey polyoma virus. SV40 tumour antigens are oncogenes. The T antigen is necessary for the cancerous state of a cell. It stimulates the host cell to replicate its DNA, binds to cellular DNA and binds to p53 protein. The polyoma virus BK and JC are human oncogenic viruses.

Adenovirus   This is highly oncogenic in animals.

Herpesviruses   They cause chromosomal damage in cells.

Epstein–Barr virus   This virus is associated with Burkitt's lymphoma and nasopharyngeal cancer.

Summary

- ✿ Cancer can be defined as the unregulated growth and production of cells. Cancer is characterized by immortalization, transformation and metastasis.
- ✿ Most of the human cancers are not hereditary and not contagious. But cancer cells have been genetically altered in one or more ways.
- ✿ Cancer is the result of a multistep process that requires several mutations.

- ✿ New mutations that arise in the clone will enhance the proliferation called clonal evolution.
- ✿ A stimulatory gene may become hyperactive and they are always dominant. These genes are called oncogenes.
- ✿ Cell division may also be activated when inhibitory genes become inactive. These inhibitory genes are recessive and are termed tumour-suppressor genes.
- ✿ DNA sequences in normal cells that closely resemble the oncogenes are proto-oncogenes. A variety of chromosomal anomalies like addition of chromosomes, loss of chromosomes and chromosomal rearrangements are the common features of most of the tumour cells.
- ✿ Carcinogens are agents that cause cancer. UV rays, X-rays and many chemicals are proven carcinogens.
- ✿ Viruses can cause malignant tumours in animals and are referred to as oncogenic viruses. There are two classes of tumour viruses—DNA tumour viruses and RNA tumour viruses.

REVIEW QUESTIONS

1. List the characteristics of a cancer cell.
2. Distinguish between benign and malignant tumours.
3. Cite an example to show that cancer is heritable.
4. What is clonal evolution with regard to cancer?
5. Explain that cancer production is a multistep process.
6. Give the various groups of genes involved in cancer formation.
7. Differentiate proto-oncogene from oncogene.
8. List the proven carcinogens.
9. What are tumour viruses?

13

CONTROL OF GENE EXPRESSION

Objectives

- ✿ To envision the concepts in gene regulation
- ✿ To understand the control of gene expression in bacteria
- ✿ To analyse the kinds of controlling mechanisms in eukaryotic gene regulation

Key Terms

operon	structural genes	promoter
operator	regulator protein	regulator gene
inducible	repressible	negative control
positive control	allosteric proteins	co-repressor
lac operon	β-galactosidase	thiogalactoside
coordinate induction	inducer	allolactose
lac mutations	partial diploid	constitutive
super repressors	*qa* gene	operator mutations
promoter mutations	glucose effect	catabolite repression
trp operon	repressible	chorismate
attenuation	attenuator	antiterminator
RNA regulators	antisense RNA	ompF
ompC	micF RNA	cluster
dehydrogenase	dehydroquinase	*GAL* gene cluster
transferase	epimerase	rinase
heat-shock response	heat-shock genes	ecdysone
heat-shock proteins	puffing pattern	imaginal discs
steroid hormone	glucocorticoids	cytotoxic
ovalbumin	5S gene	long-term regulation

autogenous	nucleosomes	ribosomal RNA
nonhistone	oogenesis	haemoglobin
informational RNA	gene amplification	blastula
cleavage	differentiated	organogenesis
histones	DNA methylation	5-methylcytosine
CpG islands	dehydrase	quinic acid
histone acetylation	maternal ribosomes	
cellular differentiation	polytene chromosome	
short-term regulation	DNase I hypersensitivity	
lampbrush chromosomes		
catabolite activator protein (CAP)		
hereditary persistence of foetal haemoglobin		

INTRODUCTION

The Central Dogma of molecular biology states that the genetic information stored in DNA, the master molecule, is transcribed into RNA and then translated into protein. This flow of information is precisely regulated. Since the proteins required by the organism vary from time to time, all the genes in an organism are not active all the time. Many proteins that are required for embryonic life are not produced in adult life. Similarly proteins that play an important role in adult life are not synthesized in embryonic life. The gene expression is precisely regulated. The mechanism of gene regulation has been extensively studied in *E. coli* and in certain higher organisms.

Regulation may be brought out at various levels during the flow of genetic information. First, regulation may be through the alteration of gene structure. Change in the chromatin structure is the major process in gene regulation. The second level is transcription. It limits the protein production early. The third potential point of gene regulation is in mRNA processing. The fourth point of control is the regulation of RNA stability. The fifth point is at the level of translation involving a number of factors, enzymes and RNA molecules. Finally, many proteins are modified after translation. Gene expression may be regulated at any or all of these points.

GENE REGULATION IN BACTERIA

The gene regulation mechanisms were first investigated in bacterial cells. The availability of mutants made it possible to unravel the mechanisms of gene regulation. The regulation mechanisms have been studied in detail in the intestinal bacteria *Escherichia coli*. It is described as regulation of gene action or genetic control of protein synthesis. The control can be exercised either by induction or by repression, either at the transcription level or at the translation level.

OPERON STRUCTURE

Many bacterial genes of related functions are grouped and are under the control of a single promoter. These genes are transcribed into a single mRNA. A group of bacterial structural genes that are transcribed together is called an **operon**. The components of an operon are

i. Structural genes, that are transcribed as a single mRNA;

ii. A promoter that controls the structural gene that lies upstream of the structural genes. RNA polymerase binds to the promoter and moves downstream transcribing the structural genes; and

iii. An operator that lies between the promoter and structural genes and the regulator protein binds to the operator.

A regulator gene though not considered as a part of the operon, affects the operon function. The regulator gene has its own promoter and transcribes a regulator protein that binds to the operator.

INDUCIBLE AND REPRESSIBLE OPERONS

Generally there are two types of transcriptional control: negative control, in which a regulatory protein acts as a repressor, binding to DNA and inhibiting transcription; and a positive control, in which a regulatory protein acts as a stimulator for initiating transcription.

In a negative inducible operon, the regulator gene produces a repressor which binds to the operator. This binding blocks the binding of the RNA polymerase enzyme with the promoter. This prevents the transcription of the structural genes. To initiate transcription, the binding of the repressor to the operator must be prevented. So an inducer is required to "turn on" the transcription as the inducer binds to the repressor and removes it from the operator site. This type of system is the inducible system. The regulatory protein has two binding sites: a site that binds to DNA and a site that binds to an inducer which is a small molecule. The shape of the repressor is altered and thereby prevents it from binding to the DNA. Proteins of this type, which change shape on binding to another molecule, are called allosteric proteins.

The structural genes are not transcribed when an inducer is absent and the repressor binds to the operator. This is an adaptive mechanism. This is because when the precursor is not available, it would be wasteful for the cell to synthesize enzymes when there is no substrate available to metabolize. As soon as the precursor is available, some of it binds to the repressor preventing it from binding to the operator. The RNA polymerase can now bind to the promoter and transcribe the structural genes. An operon with negative control regulates the synthesis of the enzyme economically as the enzyme is synthesized only when the substrate is available.

Some operons with negative control are repressible since transcription normally occurs and has to be turned off or repressed and are called the negative repressible operons (Figure 13.1). The regulator proteins are synthesized in an inactive form and cannot bind to the operator on their own.

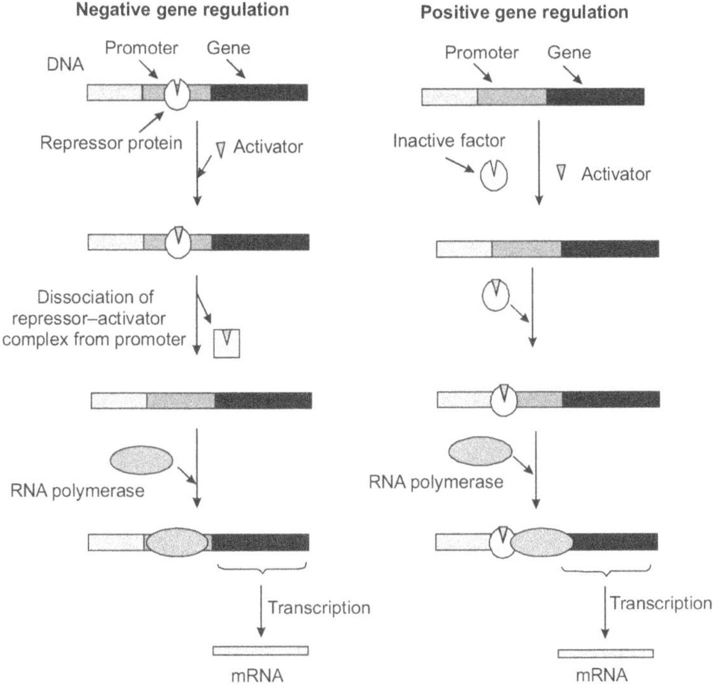

Figure 13.1   Positive and negative gene regulation
(*Source*: www.employees.csbsju.edu)

Since the repressor is not active, the RNA polymerase readily binds to the promoter, and structural genes are transcribed. The transcription is turned off by the activation of the repressor. A small molecule termed as co-repressor binds to the repressor and activates it. Like the inducible operons the repressible operons are also economical as the enzymes are synthesized only when needed. In positive control, a regulatory protein (other than operator) binds to DNA and initiates transcription. Positive control could also be inducible or repressible.

THE *lac* OPERON IN *E.COLI*

Francois Jacob and **Jacques Monod** described the "operon model" in 1961 for genetic control of lactose metabolism in *E.coli* and explained operon as the basic unit of transcriptional control. The monosaccharide lactose is found in milk as one of the major carbohydrates. Lactose is metabolized by the *E. coli* bacteria present in the gut of mammals. The transport of lactose across the cell membrane in *E. coli* is mediated by the enzyme permease. Lactose is broken down to glucose and galactose by an enzyme β-galactosidase. This enzyme can convert lactose into allolactose. Another enzyme without known specific function namely thiogalactoside transacetylase is also produced by *lac* operon (Figure 13.2).

The enzymes β-galactosidase, permease and transacetylase are encoded by the structural genes *lacZ*, *lacY* and *lacA* respectively. When lactose is not present in the medium, the levels of these enzymes are low. If lactose is added to the medium, the rate of synthesis of all the three enzymes is stimulated. This is due to the transcription of *lacZ*, *lacY* and *lacA* genes. The simultaneous synthesis of all the three enzymes is due to coordinate induction, stimulated by a specific molecule, the inducer. The lactose molecule appears to be the inducer here and allolactose is also identified as an inducer.

A common promoter namely *lacp* is responsible for the transcription of all the three genes. Upstream of the promoter is a regulator gene, *laci*, which has its own promoter.

The repressor protein is produced by the *laci* gene. The repressor is made of four identical polypeptides with two binding sites, one for the inducer and the other for the DNA. The repressor binds to the operator, *laco* in the absence of lactose.

The RNA polymerase binds to the promoter *lac p* which is upstream of the structural genes. The structural genes are transcribed when the RNA polymerase binds to the promoter and moves down the DNA. When the repressor is bound to the operator, the RNA polymerase binding is blocked and transcription is prevented. On the other hand when lactose is present (some of it is converted to allolactose), it binds to the repressor and the repressor is released from the DNA. Thus the repressor is inactivated and the binding of RNA polymerase to the promoter initiates transcription of *lacZ*, *lacY* and *lacA*.

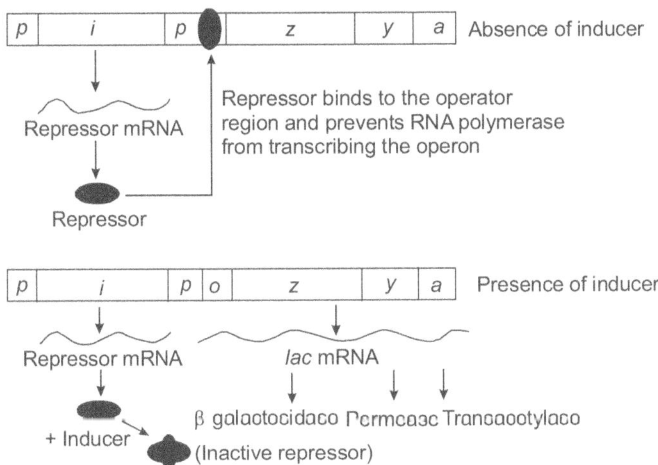

Figure 13.2   The *lac* operon
(*Source*: www.dwb.uni.edu)

β-galactosidase converts some of the lactose into allolactose. Allolactose attaches to the repressor and alters its shape so that the binding of the repressor to the operator is blocked and the operator site is free. Some compounds related to allolactose can also act as inducers. **Isopropylthiogalactoside** (IPTG) is one such inducer that inactivates the repressor and permits transcription of the structural genes.

lac Mutations

The structure and function of *lac* operon can be studied by analysing the mutations of *lac* operon. Jacob and Monod worked on the partial diploid strains of *E. coli* carrying two copies of *lac* operon obtained by recombination. Partial diploids with the structural gene mutations *lacZ⁺ lacy⁻* on the bacterial chromosome and *lacZ⁻ lacy⁺* on the plasmid functioned normally producing β-galactosidase and permease in the presence of lactose (the genotype is *lacZ⁺ lacy⁻/lacZ⁻ lacy⁺*). This observation shows that a functional β-galactosidase gene *lacZ⁺* is enough to produce the enzyme β-galactosidase. The same is true for *lacY⁺* gene.

Jacob and Monod isolated regulator-gene mutations that affected enzyme regulation. These mutations affected the β-galactosidase and permease production. Some mutations are constitutive; these mutations cause lac enzymes to be produced all the time without depending on the presence or absence of lactose. The mutation of the regulator gene is designated as *lacI⁻*. It was demonstrated that *lacI⁺* gene is dominant over the *lacI⁻* gene. A single copy of *lacI⁺* (genotype *lacI⁺/lacI⁻*) is sufficient to produce the enzymes. A partial diploid with genotype *lacI⁺ lacZ⁻/lacI⁻ lacZ⁺* functioned normally, and the enzyme β-galactosidase was synthesized only when lactose was present. In this organism the *lacI⁺* gene on the bacterial chromosome was functional, but the *lacZ⁻* gene was defective; but on the plasmid, the *lacI⁻* gene was defective, but the *lacZ⁺* gene was functional. This shows that the *lacI*+ gene could regulate a *lacZ*+ gene located on a different DNA molecule, indicating that the *lacI*+ gene product was able to diffuse to either the plasmid or the chromosome.

Some *lacI* mutations prevented transcription even in the presence of inducers like lactose and IPTG. These mutations were denoted as super repressors [*lacI^s*]. The *lacI^s* mutations produced a repressor where the inducer-binding site is altered preventing the binding of the inducer to the repressor, thereby the repressor could attach to the operator and prevent transcription of the *lac* genes. The *lacI^s* mutations were dominant over *lacI*+.

The operator mutations of constitutive category occurred at the operator site and denoted as *lacO^c*. The *lacO^c* mutations altered the sequence of DNA at operator preventing the repressor binding. The *lacO^c* mutation is dominant over *lacO*+.

Promoter mutations (*lacP*−) occur at the promoter site and they interfere with the RNA polymerase binding to the promoter. *E.coli* with *lacP*− mutations does not produce enzymes either in the presence or in the absence of lactose.

POSITIVE CONTROL AND CATABOLITE REPRESSION

Bacteria metabolize glucose in the presence of other sugars. When glucose is present, the genes that are involved in the metabolism of other sugars are repressed. This phenomenon is known as catabolite repression or glucose effect. For example, the *lac* operon is efficiently transcribed only if lactose is present and glucose is absent.

Catabolite repression occurs due to positive control in response to glucose. This is achieved by the binding of a protein dimer namely the catabolite activator protein (CAP) to the site that is located upstream of the promoter of the *lac* genes.

RNA polymerase binds to promoter only when CAP is bound to DNA. The CAP forms a complex with cyclic AMP (cAMP) before it binds to DNA. A high concentration of glucose within the cell lowers the amount of cAMP, and so little cAMP–CAP complex is available to bind to the DNA. So RNA polymerase cannot bind efficiently to promoter. But low levels of glucose stimulate high levels of cAMP leading to an increased cAMP–CAP binding to DNA (Figure 13.3). This enhances the binding of RNA polymerase to the promoter and increases the transcription of *lac* genes by 50-fold.

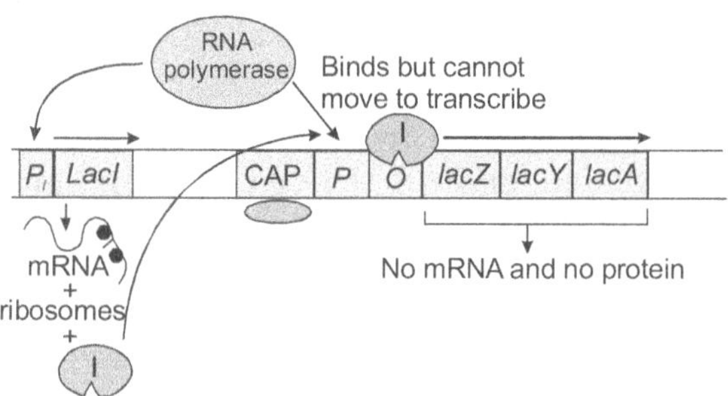

Figure 13.3   Catabolite repression in *E. coli*
(*Source*: www.biology.kenyon.edu)

trp OPERON OF *E. COLI*

The *lac* operon is an example for inducible operon as normally transcription does not take place and it must be turned on. In contrast some operons are **repressible**; as the transcription is normally turned on and must be repressed. An example for a repressible operon is the tryptophan (*trp*) operon in *E.coli* controlling the biosynthesis of the amino acid, tryptophan.

There are five structural genes in *trp* operon namely *trpE, trpD, trpC, trpB* and *trpA*. They produce components of three enzymes that convert chorismate into tryptophan. The first structural gene *trpE* contains 5'UTR (the long 5'untranslated region) that does not code for any of these enzymes but binds to the promoter and transcribes the five structural genes into a single mRNA that is translated into enzymes that convert chorismate into tryptophan.

The regulator gene, *trpR,* encodes a repressor that alone cannot bind DNA. The *trp* repressor has two binding sites, one that binds to DNA at the operator and another that binds to the activator, the tryptophan. The tryptophan binding makes a conformational change in the repressor that enhances the binding to the operator. When the operator is occupied by the tryptophan repressor, RNA polymerase cannot bind to the promoter and so there is no transcription. So, when tryptophan level is low in the cells, the *trp* operon transcription takes place and tryptophan is synthesized. On the other hand, when tryptophan level is high, transcription is inhibited preventing tryptophan synthesis.

ATTENUATION

The continuation of transcription is controlled by other factors too. In attenuation, transcription begins at the start site but termination occurs prematurely. In *trp* operon the repression is not the only method of regulation. In 1970, **Charles Yanofsky** and his colleagues observed attenuation as the other regulatory mechanism in *trp* operon. In *trp* operon the 5′ UTR contains four regions that allow the folding of 5′UTR into two different secondary structures. One of the secondary structures contains two hairpin structures with a string of uracil nucleotides at the end. This is the terminator and is called an attenuator. This attenuator structure is produced when tryptophan level of the cells is high and causes termination of transcription prematurely before the *trp* structural genes are transcribed.

Another alternative structure of the 5′UTR is also a hairpin but without a string of uracil bases and does not function as a terminator. When cellular level of the tryptophan is low, this structure is formed and *trp* gene transcription is not terminated. RNA polymerase continues past the 5′ UTR into the structural genes-coding region, the enzymes required for tryptophan synthesis are produced. As it prevents the termination of transcription, it is called as an antiterminator.

The *E. coli trp* operon has dual control namely repression and attenuation. Repression reduces transcription but attenuation further reduces transcription, and these two together reduce transcription of the *trp* operon by 600-fold. The attenuation and repression respond to different signals: repression depends on the cellular levels of tryptophan, whereas attenuation responds to the number of tRNAs charged with tryptophan (Figure 13.4).

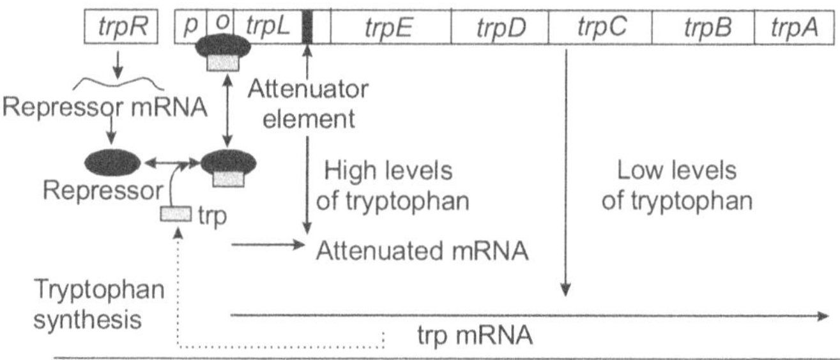

Figure 13.4   The *trp* operon
(*Source*: www.dwb.uni.edu)

RNA REGULATORS—ANTISENSE RNA IN GENE REGULATION

RNA regulators are small RNA molecules that are complementary to specific sequences in mRNAs and are termed antisense RNA. They are involved in control of gene expression by binding to RNA sequences and inhibiting translation.

An example for antisense RNA regulation is *ompF* gene of *E.coli*. *ompF* and *ompC* are the genes in *E. coli* responsible for producing outer membrane proteins that act as diffusion pores. These two genes are translated in most conditions. When the osmolarity of the medium increases, a regulator gene *micF* (mRNA-interfering complementary RNA) is activated and micF RNA is produced. micF RNA is an antisense RNA that binds to a complementary sequence in the ompF mRNA and inhibits the ribosome binding. Due to this, translation is inhibited and few ompF proteins are produced in the outer membrane thereby reducing the movement of substances across the membrane.

EUKARYOTIC GENE REGULATION

Gene regulation in bacterial and eukaryotic cells, are mostly similar with some differences. Eukaryotic genes are not organized into operons. Each structural gene has its own promoter and so it is transcribed separately and are not transcribed together. The chromatin structure is complicated and the DNA must unwind from the histone proteins before transcription and there are diverse mechanisms involved in eukaryotic gene regulation. In eukaryotes there are two kinds of gene regulation namely reversible or short-term gene regulation and irreversible or long-term gene regulation.

SHORT-TERM REGULATION

This kind of regulation occurs as response to changes in the environment involving changes in the enzyme or hormone activity.

i. *The qa gene cluster in Neurospora*
Neurospora contains enzymes necessary for the synthesis of amino acids like tyrosine and phenylalanine which are aromatic molecules. When such aromatic compounds are produced at a high level, synthesis of scavenging enzymes are

induced that break them down to non-aromatic compounds. The three induced enzymes are a dehydroquinase, a dehydrogenase and a dehydrase encoded by a cluster of genes *qa*-2, *qa*-3 and *qa*-4. These are inducible genes. Quinic acid is an aromatic metabolite that is responsible for the induction of these enzymes.

The *qa* genes are under positive control. The inducer quinic acid combines with a regulator to turn on *qa* gene expression. Thus the expression of *qa* gene cluster is regulated by the *qa*-1 gene product.

ii. *GAL gene cluster in yeast* Three enzymes are involved in galactose fermentation in yeast. The enzymes are a rinase, a transferase and an epimerase. They are encoded by closely linked genes *GAL*1, *GAL*7 and *GAL*10. In yeast, the *GAL* cluster is controlled by the product of a different unlinked gene '*i*'. The i repressor of yeast acts not on the *GAL* gene cluster directly but on another unlinked gene, *GAL*4. The *GAL*4 encodes a positive regulator protein, like the qa-1 protein that directly stimulates transcription in the GAL gene cluster. The *GAL*4 in turn is negatively controlled by the '*i*' gene product.

iii. *Heat-shock response* In *Drosophila*, exposure to high temperature or to a variety of metabolic inhibitions stops the transcription of most of the active genes; instead some heat-shock genes are transcribed. New heat-shock proteins are synthesized. The total RNA synthesis drops by 70% and 40% of the total protein synthesis is devoted to heat-shock protein synthesis.

Heat-shock response in *Drosophila* is observed by increasing the temperature from 25° to 37°C. The response was first reported in larval tissues as a heat-induced change in the puffing pattern of salivary gland chromosomes. Puffing of a band in a polytene chromosome indicates higher transcriptional activity. **F. Ritossa** noticed that within one minute after the temperature was raised, most already existing puffs began to disappear and nine new puffs were detected. The heat-shock response was observed in two bands 87A and 87C that reside in the right arm of chromosome 3 and the rest in other chromosomes. In *Drosophila* the heat causes induction of nine cluster gene loci and several minutes later, there is translation of these transcripts into proteins.

Experiments indicate that stimulation of gene transcription is the primary response to heat-shock. The secondary response is the synthesis of heat-shock proteins and their involvement in protecting the cell from damage.

iv. *Hormonal regulation*   In 1960, **U.Clever** and his co-workers reported hormonal effect at the chromosomal level in the dipteran, *Chironomus tentans*. Clever observed puffing pattern in the development of chironomus larvae and established that certain bands puff at one stage and regress at a later stage. He injected a steroid hormone ecdysone into non-moulting larvae and noted puffing of band I-19A within 10 to 15 minutes, I-19A puff regressed and band I-18C began to puff followed by band IV-2B. This puffing sequence is similar to that which occurs before moulting when the insects release ecdysone hormone into the haemolymph. The experiments by Clever indicate that ecdysone "turns on" the transcription of particular genes.

Similar experiments were performed by **M.Ashburner** in *Drosophila*. He isolated the salivary glands from third-instar larvae, cultured them *in vitro* and added ecdysone at specific intervals. Six bands puffed rapidly and bands 23E and 74E showed puffing within five minutes, reached a maximum size and regressed within four hours. The early puffs were unaffected by protein inhibitors but regressed prematurely if ecdysone was washed off from the culture medium during the four-hour period. About 100 late puffs were observed three to ten hours after exposure to the hormone.

Late puffing does not occur in the presence of protein inhibitors. It was concluded that late puffs are stimulated by some early puff gene products and not by ecdysone alone. Early puffing is the primary response to the hormone and late puffing is the secondary response. In heat-shock response the stimulation is stress but here it is a hormone normally produced in the larvae.

It was demonstrated by **O. Pongs** and his colleagues in *Drosophila* that ecdysone stimulates gene expression at the level of the chromosome. **J.J. Bonher** and **M.L. Pardue** conducted experiments to find out the specificity of the genes involved. They removed the imaginal discs from the developing larvae, cultured them and exposed them to ecdysone 2 ^{3}H-uridine for different lengths of time. The ^{3}H-mRNA from the cells was then exposed to salivary gland chromosome preparations under *in situ* hybridization conditions. The localization of tritium grains on polytene bands reveal that these bands are not the same bands that undergo puffing in larval tissues. The results indicate that a single hormone activates one set of genes in one cell type and a second set of genes in another cell type. This difference is brought out by tissue-specific receptor proteins that combine with ecdysone and recognize distinct controlling elements. The other possibility is that the cell type might modify its controlling elements that the hormone recognizes a different set of genes.

Steroid hormone and gene transcription

Steroid hormones enter target cells and accumulate in the nucleus. The transport of the steroid hormone to the nucleus is carried out by specific steroid receptors in the cytoplasm. The receptor-binding causes changes in the receptor shape and migration to the nucleus, a primary response. Steroids act as effectors and the receptor proteins act as regulator molecules and get activated by binding to the effector.

C.Tomkins, **K.Yamamotto**, **C.Sidley** and associates conducted experiments with mouse lymphoma cell line S49. The cells were rapidly killed by adding glucocorticoids and mutant cell lines were isolated that are resistant to the cytotoxic effects of the steroid. The mutants were produced at various stages of steroid utilization. Most of the mutants were found to be receptor-deficient (r⁻), where the labelled steroid enters the cells but is not bound to any cytoplasmic or nuclear protein. Some mutant lines transport steroid to the nucleus with reduced efficiency (nt⁻), and others showed excessive binding of steroids by the nucleus (ntⁱ). The receptor proteins of nt⁻ and ntⁱ showed altered molecular properties. It was proposed that some nt⁻ and ntⁱ mutations mask the structural genes or genes for steroid-receptor proteins.

The steroid–receptor complex which enters the nucleus, stimulates the specific gene transcription. R.Palmiter's experiment involving oestrogen administration into chicken oviduct cells showed the specific synthesis of pre-mRNA transcripts for ovalbumin, the egg-specific protein. The hormone–receptor complexes do not act indirectly as stimulating pre-mRNA to mRNA processing or transport of ovalbumin mRNA to ribosomes but have direct effect on gene expression.

Mechanisms of Short-term Regulation

There are proteins that bind to target genes, and proteins that bind to RNA polymerases thereby altering the initiation or termination of transcription. In *Xenopus laevis*, **R. Roder** and his colleagues discovered that the 5S gene possesses an internal promoter and transcribes RNA polymerase III. The transcription depends on the presence of a 40,000 dalton protein factor that binds to the internal promoter facilitating the polymerase binding. The factor also binds to 5S RNA itself and cannot stimulate transcription of the gene. Thus the

5S genes appear to be autogenous regulators. A model is proposed to explain the system. Cells contain factors that bind to 5S RNA and are available for incorporation into ribosomes. When ribosomes assemble, the 5S factor is released, permitting the binding to 5S gene and synthesizes more 5S RNA. The new 5S RNA binds to the factor and transcription is turned off; the factor-bound 5S RNA migrates to the ribosome.

Eukaryotic DNA is organized in such a way that it wraps around nucleosomes and the nonhistone proteins may also alter the DNA conformation. The DNA can be transcribed only when it is exposed to polymerase enzymes. So, certain physical conformations may prevent gene expression and others may promote it. Some short-term regulatory molecules may act to alter the physical state of the genes.

Long-term Regulation

i. *Expression of haemoglobin genes* In humans, haemoglobin genes are subjected to two kinds of regulation. In the first one the entire haemoglobin gene family is transcribed only in the erythroid cells. On the other hand, the second operates on individual genes that enable the globin synthesis at specific stages of development.

Differential expression of globin genes is established in the development. In embryonic condition ε (epsilon) and ξ (zeta) globin chains are synthesized. In the foetus only α (alpha) and γ (gamma) globin chains of foetal haemoglobin (HbF) are synthesized; but in newborn β (beta) and δ (delta) globin chains are produced and continue to be in adult life as adult haemoglobin (HbA). The foetal $\rightarrow$ adult change in human Hb is altered in genetic disorders called HbFH (hereditary persistence of foetal haemoglobin). At 32–34 weeks of gestation, generally the γ genes will be turned off and β and γ genes are activated. But in HbFH, β- and

δ-globins are not synthesized and γ-globin is synthesized even after birth. This indicates the involvement of controlling element in γ gene expression.

ii. *Gene regulation during oogenesis*   In amphibians and echinoderms, egg maturation involves transcription of large amount of mRNA and rRNA. In amphibians the oocyte possesses "lampbrush" chromosomes. In *Xenopus,* in the diplotene stage of prophase I, the oocyte lampbrush chromosomes transcribe large amount of mRNA that may produce 40,000 different kinds of proteins. But most of the mRNA is not translated and take up regulatory function. The mRNA of the lampbrush chromosome is transcribed selectively and remains for a long time throughout egg maturation. When the egg is fertilized and the zygote starts to divide, the lampbrush mRNA directs the development. The eggs of insects, echinoderms and amphibians accumulate mRNA or informational RNA well in advance and utilize it during development. Similar to mRNA, the ribosomal RNA too is synthesized and stored in the amphibian for future development. At the pachytene stage of prophase I, there is rapid synthesis of rRNA and this is stored in the egg as maternal ribosomes. During oogenesis the rRNA genes undergo repeated replication when the rest of the gene amplification is restricted to a particular period of development and its mechanism is not understood.

iii. *Cellular differentiation*   The fertilized egg undergoes cleavage and forms blastula when the cells are similar. The individual cells are determined to develop into specific cell types like liver, heart, etc. Later these determined cells get differentiated into specialized forms. There is a sequence of events occurring one after another during differentiation. Differentiation begins during later blastulation and continues throughout in organogenesis. During embryonic development, the cells differentiate but their DNA content remains unaltered. There is preferential synthesis of some proteins. For example, in erythrocytes the haemoglobin and in muscle the actin and myosin are synthesized specifically. It is suggested that when some determined genes are turned on, certain proteins are synthesized. Some of these proteins may act in the nucleus to sustain transcription of certain genes and others act in the cytoplasm to sustain the stability of particular mRNAs or to promote synthesis of specific proteins. In a differentiated cell the cytoplasm thus accumulates regulatory proteins that reinforce the differentiated condition.

iv. *DNase I hypersensitivity*   This is a type of increase in the sensitivity of chromatin to degradation by DNAse I enzyme. The DNA when tightly bound to histones is not accessible to the enzyme DNase I digestion. On the other hand when the DNA is less tightly bound to histones, it is easily degraded by the DNase enzyme. Thus the DNA–histone association is indicated by the DNase I enzyme activity.

When the genes become active, the regions around the genes become sensitive to the enzyme and these regions are called DNase I hypersensitive sites. In these regions the chromatin structure is more relaxed and serves as the binding site for regulatory proteins and the genes are actively transcribed.

v. *DNA methylation*   The methylation of cytosine bases results in 5-methylcytosine. When DNA is methylated there is inhibition of transcription in vertebrates. The cytosine bases that are closer to guanine are often methylated and it occurs in the same strand (CpG). The DNA segment with many CpG sequences is termed CpG islands. When genes are not transcribed, there is methylation of CpG islands. At the start of transcription, the methyl groups are removed. CpG methylation is seen in long-term inactivation of X chromosome in female mammals.

vi. *Histone acetylation*   Chromatin structure can be altered by addition of acetyl groups to histone proteins. This is mediated by acetyl transferase enzyme. By acetylation the DNA is separated from the histones. Some transcription factors have acetyl transferase activity and regulate transcription.

Summary

- In bacteria, the control of gene expression is carried out by turning genes on and off in response to the environment.

- A group of bacterial structural genes that are transcribed together is called an operon.

- The components of an operon are structural genes, promoter and operator.

- A regulator gene affects the operon function.

- Generally there are two types of transcriptional control; a negative control and a positive control.

- The *lac* operon of *E.coli* controls the transcription of three genes governing the lactose metabolism.

- It is an inducible operon. A positive control in the *lac* operon is catabolite repression.

- The structure and function of *lac* operon was worked out by analysing the different kinds of *lac* mutations.

- The *trp* operon is a repressible operon controlling the synthesis of tryptophan. In attenuation, transcription is initiated but terminated prematurely.

- Apart from proteins, there are RNA regulators that control gene expression and are termed antisense RNA.

- In eukaryotes there are two kinds of gene regulation namely reversible or short-term gene regulation and irreversible or long-term gene regulation.

REVIEW QUESTIONS

1. How can inducible and repressible operons of bacteria be distinguished?

2. In the *lac* operon of *E. coli,* what is the function of each of the following?

   i.   regulator

   ii. operator

   iii. promoter

   iv. structural genes Z, A and Y

3. What is the biological significance of catabolite repression?

4. Citing examples, distinguish between positive regulation and negative regulation.

5. List out the regulatory processes that modulate the *trp* operon.

6. What is antisense RNA? How does it regulate gene expression?

7. What changes take place in chromatin structure to regulate eukaryotic gene expression?

8. Discuss the experimental evidences in support of short-term gene regulation in eukaryotes.

9. What are the diverse mechanisms observed in long-term gene regulation with reference to eukaryotes?

14

DEVELOPMENTAL GENETICS

Objectives

- ✿ To study the coordinated activity of genes during development
- ✿ To observe the patterns of normal development in eukaryotes
- ✿ To analyse the genetic control of development

Key Terms

totipotent	caspases	*egg-polarity genes*
morphogen	dorsal protein	cactus
toll	twist	*decapentaplegic*
bicoid	oskar	exuperentia
nanos	gap genes	*krup-pel*
pair-rule gene	even-skipped	runt
hairy	*fushi tarazu* gene	antennapedia
necrosis	apoptosis	procaspase
gooseberry gene	*homeotic* genes	cloning experiments
segment polarity genes	developmental ground state	
homeotic complex BX–C	shibire temperature-sensitive 1	

INTRODUCTION

Genes control traits at different levels like phenotypic, biochemical and molecular levels. The individual gene functions through the synthesis of protein and its regulation. However, importance should be given to the coordinated activity of different genes and the regulation of these activities. The activities of genes lead to different developmental patterns, starting from the egg or zygote to the fully developed adult. The development process is so precise, that each step of development is dependent on the preceding stage. Any disruption in the normal sequence of events results in an abnormal phenotype, sometimes due to mutations.

It is well known that every cell of an organism, at any stage of its development, contains the entire genome with all its genes. Then, how are these genes regulated precisely to bring out the complexity in the organisms? This problem can be analysed in this chapter by exploring the mechanisms involved in different organisms.

Every organism begins life as a unicellular, fertilized egg. This single-celled zygote undergoes repeated cell division and develops into a complete organism. Initially each cell in the embryo is totipotent—it has the potential to develop into any cell type. All cells of an organism are identical in their genetic content and the development proceeds by differential expression of these genes.

CLONING EXPERIMENTS

The totipotency is explained to show that all cells are similar. Then how do different cell types arise? One possibility is that, throughout development, genes might be selectively lost or altered, causing different cell types to have different genomes. The other possibility is that each cell might contain the same genetic information, but different genes might be expressed in each cell type. Cloning experiments demonstrated, that in most animals the complete set of genetic information is retained during development. In 1952, Robert Briggs and Thomas King removed the nuclei from unfertilized oocytes of the frog *Rana pipiens*. They then isolated the nuclei from frog blastula (early embryonic stage) and injected these nuclei individually into the oocytes. These eggs developed into a complete frog. In the 1960s, John Gurdon used these methods to successfully clone few frogs with nuclei isolated from intestinal cells of tadpoles. In 1997, Ian Wilmut and his co-workers from Roslin Institute of Scotland cloned Dolly, the sheep, by using the genetic material from a differentiated cell of an adult animal. They fused an udder cell from a white-faced Finn Dorset ewe with an enucleated egg cell and stimulated the egg electrically to initiate development. After a week, these embryos were implanted into the Scottish black-faced surrogate mother.

Dolly, the first mammal cloned from an adult cell, was born on July 5, 1996. Following sheep, mice, calves, cats, dogs and other animals have been cloned from differentiated adult cells. These cloning experiments demonstrate that during development, the genetic material is not lost or altered but there is selective expression of genes. In recent years, powerful genetic and molecular techniques have developed to probe the processes of development. *Drosophila* is used as a model system to study the genetics of development.

GENETICS OF *DROSOPHILA* DEVELOPMENT

One aspect of development and differentiation is the differential activation of various genes at different stages of development. Mutations in specific genes lead to failure of the development of specific organs. These mutations provide information about gene control of development. Some examples are given below.

Insect bodies consist of serially repeating units, which differentiate into specific structures and patterns according to their position. In many insects, segments are generated sequentially as the cells of the embryo proliferate. By contrast in *Drosophila,* the entire body plan is established simultaneously at the blastoderm stage. The adult fly consists of

1.   head

2.   three thoracic segments—prothorax, mesothorax and metathorax, designated as T_1, T_2 and T_3. T_1 carries the first pair of legs; T_2 carries the wings and the second pair of legs and T_3 carries the third pair of legs and the halteres (rudimentary second pair of wings).

3.   Eight abdominals (AB_1 to AB_8).

These segments can be seen even in the larval stage.

In *Drosophila,* after fertilization the diploid nucleus of the embryo divides nine times without the division of the cytoplasm resulting in a single multinucleate cell (Figure 14.1).

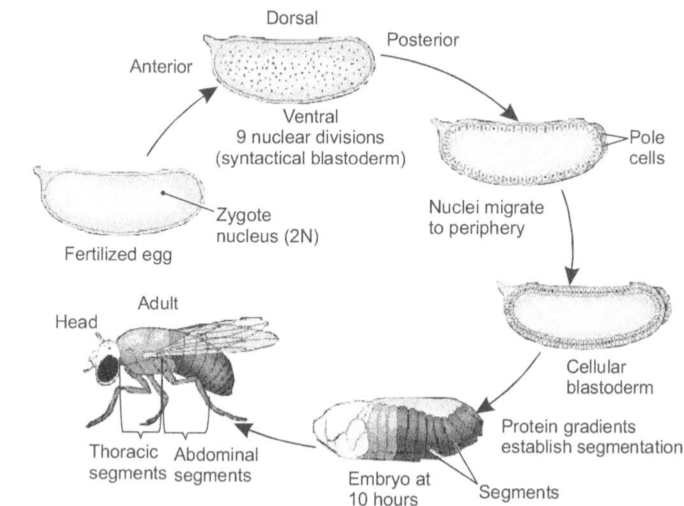

Figure 14.1   Life cycle of *Drosophila*
(*Source*: www.biology.kenyon.edu)

In the beginning these nuclei are scattered in the cytoplasm and then migrate towards the periphery of the embryo and divide again several times. The cell membrane grows and creates a layer of approximately 6000 cells. At one end of the embryo, four nuclei develop into pole cells giving rise to germ cells. Further development occurs in three stages:

1.   The anterior–posterior axis and the dorsal–ventral axis of the embryo are established.

2.   The orientation and the number of the body segments are determined.

3.   The identity of each segment is established. These three stages are controlled by different sets of genes.

Genes Involved in the Development

Egg-polarity genes   The two main axes of development are determined by the *egg-polarity* genes. There are two sets of genes: one determines the anterior–posterior axis and the other determines the dorsal–ventral axis. These genes produce morphogen, a protein which sets up a concentration gradient that affects the developmental fate of the surrounding region.

In the maternal parent, these *egg-polarity* genes are transcribed into mRNAs during egg formation. The mRNAs are translated into proteins after fertilization. These proteins are responsible for the two main axes of the embryo.

Genes determining the dorso-ventral axis   In the fly, the dorsal–ventral axis determines the back and abdomen. There are twelve different genes involved and one of the important genes is called *dorsal*. In a newly laid egg, mRNA and the protein encoded by the *dorsal gene* are evenly distributed throughout the cytoplasm but after the nuclei migrate to the periphery of the embryo, the dorsal protein becomes redistributed. On one side of the embryo, the dorsal protein remains in the cytoplasm and this side becomes the dorsal surface. Along the other side, the dorsal protein is taken up by the nuclei; this side will become the ventral surface. There are two other proteins called Cactus and Toll that regulate the nuclear distribution of the dorsal protein, which in turn determines the dorso-ventral axis of the embryo. High nuclear concentration of dorsal protein activates a gene called *twist*, which causes the mesoderm to develop. Low concentrations of dorsal protein activates a gene called *decapentaplegic*, which specifies dorsal structures. The dorsal and ventral sides of the fly are thus determined.

Genes determining antero-posterior axis   The antero-posterior pattern is mainly established by the activity of *bicoid* (*bcd*) gene and that of genes of *Oskar* group. The gene *bcd*+ is maternally expressed in the ovary in specialized cells, which form a cluster around the future anterior pole of the developing oocyte. As *bicoid* mRNA enters oocyte, it is trapped by components encoded by the genes *swallow* (*swa*) and *exuperentia* (*exu*), and thus becomes localized at the posterior end of the egg. Thus the egg gets polarized at the morphological level as well as at the molecular level. Mutants of *bcd* gene produce eggs that develop into embryos without head or thorax. Similarly, mutants for any *osk* group gene develop normal head and thoracic segments but lack entire abdomen. After fertilization, *bicoid* mRNA moves backward and *osk* RNA moves forward to about the middle of the zygote, each establishing a gradient. Subsequently, these maternally derived products bind with DNA and thus interact with the zygotic genome (Figure 14.2).

The development of the antero-posterior axis is also greatly influenced by a gene called *nanos,* an egg-polarity gene that acts at the posterior end of the axis. The *nanos* gene is transcribed in the adult female, and the resulting mRNA becomes localized at the posterior end of the egg. After fertilization, *nanos* mRNA is translated into Nanos protein, which diffuses slowly into the anterior end. Nanos protein inhibits the formation of anterior structures by repressing the translation of *hunchback* mRNA. The synthesis of Hunchback protein is therefore stimulated at the anterior end of the embryo by Bicoid protein and is repressed at the posterior end by Nanos protein. This combined stimulation and repression results in a Hunchback protein concentration gradient along the antero-posterior axis. This gradient, in turn affects the expression of other genes and determines the anterior and posterior structures.

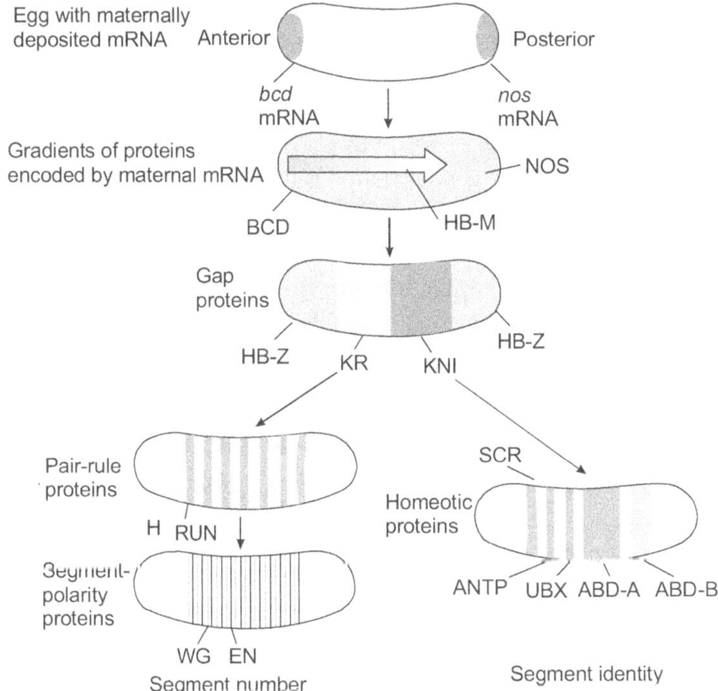

Figure 14.2   Genes in *Drosophila* for antero-posterior axis (*Source:* www.mun.ca)

Segmentation genes Segmentation genes control the differentiation of the embryo into individual segments. These genes affect the number and organization of the segments; and mutations in them usually disrupt whole sets of segments. The segmentation genes fall into three groups.

i. *Gap* genes define large sections of the embryo. The *gap* genes when mutated, created gaps in antero-posterior pattern. Mutations in the *krup-pel* gene cause the absence of several adjacent segments.

ii. *Pair-rule* genes define regional sections of the embryo and affect alternate segments. Mutations in the even-skipped gene causes the deletion of even-numbered segments, whereas mutations in the *fushi tarazu* gene cause the absence of odd-numbered segments. There are eight pair-rule genes, of which two genes *runt* and *hairy*, appear to have major function in the generation of striped patterns. These genes regulate expression of each other. By the last interphase, the transcripts of these genes generate striped patterns, a key event in development.

iii. *Segment polarity* genes affect the organization of segments. Mutations in these genes cause part of each segment to be deleted and replaced by a mirror image of part or all of an adjacent segment. For example, mutations in the *gooseberry* gene cause the posterior half of each segment to be replaced by the anterior half of an adjacent segment (Figure 14.3).

The gap genes, pair-rule genes and segment polarity genes act sequentially, affecting smaller regions of the embryo. First, the egg-polarity genes activate or repress the gap genes, which divide the embryo into broad regions. The gap genes, in turn, regulate the pair-rule genes which affect the development of pairs of segments. Finally, the pair-rule genes influence the

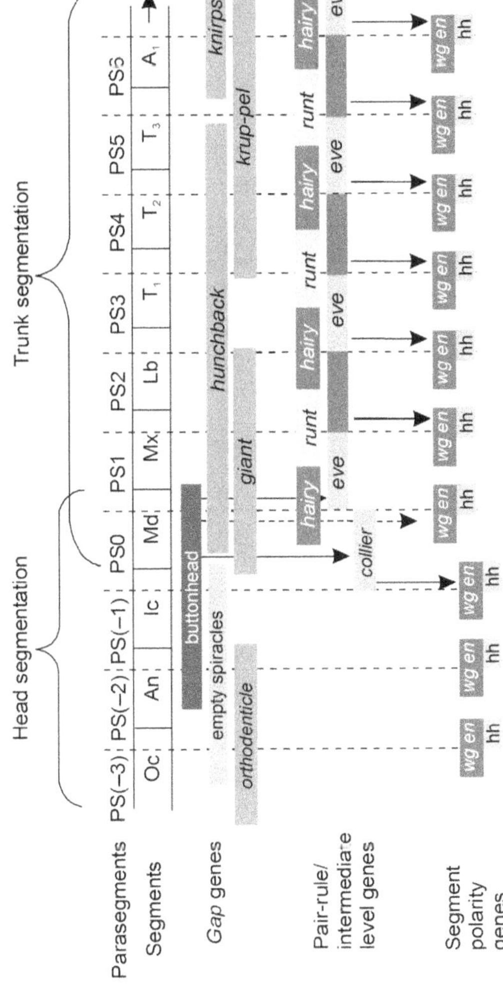

Figure 14.3 Segmentation genes involved in the development of *Drosophila* (*Source*: www.pharyungula.org)

segment polarity genes, which direct the development of individual segments.

Homeotic genes   The identity of individual segment is determined by *homeotic* genes. For example, the eyes normally arise on the head segment while the legs in the thoracic segments. But mutations in these genes can cause body parts to appear in the wrong segments. In 1940, Edward Lewis observed homeotic mutations in *Drosophila*. He observed mutations in the *antennapedia* gene, in which the legs developed in the head. Genes in ANT-C affect head and thorax. A dominant mutation converts part of the antenna into leg structures. This replacement of parts is position-specific, that is, a given internal segment is always replaced by a specific part of the leg. These observations suggest that there is some common overall developmental plan, on which the specific details of leg or antenna development are overlaid.

There are two major groups of homeotic genes in *Drosophila*. One group, the *antennapedia* complex, affects the development of the adult fly's head and anterior thoracic segments. The other group consists of the *bithorax* complex that influences the adult fly's posterior thoracic and abdominal segments. These two groups together are called the **homeotic complex** (HOM-C). The *bithorax* complex contains three genes and the *antennapedia* complex has five.

Mutation in a complex locus, the bithorax complex (BX-C), in *Drosophila* is considered to be the best example for the study of development leading to the production of adult fly. The most important mutations in BX-C are in bx^+ and bx^1. For instance, a strong bx mutation causes the anterior part of T_3 to develop as the anterior part of T_2. This suggests that bx^+ at some stage becomes active in some cells leading to their differentiation in the anterior part of T_3, so that inactivation of bx^+ remains active in anterior of T_3, but turned off in anterior of T_2. A deletion in entire BX-C is lethal and all segments (T_1 to T_3 and AB_1 to AB_8) develop into those like T_2 with its leg and wing, a state which is considered to be the "developmental ground state".

Different components of BX-C at the DNA level are required to direct differentiation of the posterior segments. A recessive mutation affecting a particular segment, transforms it to a more anterior segment, which is the ground state for this transformed element. There are also dominant mutations which transform specific segments into structures like those of segments further posterior to them.

TEMPERATURE-SENSITIVE MUTANTS

Temperature-sensitive (ts) mutants have been utilized for the study of development. These mutants may be grown for some time under permissive temperatures and then shifted to restrictive temperatures at specific developmental stages, to study the effect on the phenotype. In *Drosophila,* there is an allele named **shibire temperature-sensitive 1 (shi^{ts1})** in Japanese meaning paralysis. The mutants with this allele *shi^{ts1}* develop normally at 22°C, but when transferred to 29°C they become paralysed and when shifted back to 22°C, the flies recover mobility within minutes. Cultures shifted from 22°C to 29°C at different developmental stages revealed a number of lethal phases. A range of phenotypic defects in bristles, hair and eyes was found to be induced at different times, suggesting the sequence or order in which the developmental stages proceed. These studies indicate that the temperature-sensitive periods occur during the third instar stage of larval development. After this period the course of development is not affected by mutations.

CONTROL OF DEVELOPMENT

Studies indicate that development is a complex process consisting of a sequence of events. This process is regulated by a large number of genes. In *Drosophila,* the dorso-ventral axis and the anterior–posterior axis are determined by maternal genes. These genes encode proteins and mRNAs that are localized in the specific regions within the egg and cause specific genes to be expressed in different regions of the embryo. The proteins of these genes then stimulate the other genes, which in turn stimulate yet other genes in a cascade of control. The gene products in the cascade are regulatory proteins, which activate other genes.

In development, the regions of the embryo are determined successively (Figure 14.4). First the major axes and regions

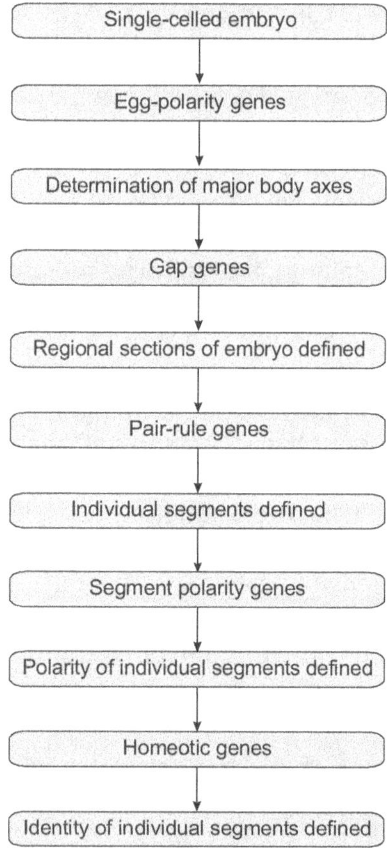

Figure 14.4   A cascade of gene regulation

are determined by polarity genes. Next the patterns within each region are determined by segmentation genes: the gap genes define large sections; the pair-rule genes define regional sections and affect alternate segments; and the segment polarity genes affect individual segments. Finally the homeotic genes provide unique identity to each segment.

PROGRAMMED CELL DEATH IN DEVELOPMENT

Generally, cells have a limited lifespan and they die and are continuously replaced. In development, many parts are shaped due to cell death. An example is the tadpole's tail during metamorphosis. Cell death is an integral part of life and is used to eliminate dangerous cells that go out of control of normal development.

Cell death is initiated by the cell itself and is termed **apoptosis**, or cellular suicide (Figure 14.5). In apoptosis the nucleus and cytoplasm shrink; DNA is degraded and the cells are engulfed by neighbouring cells without leakage of its contents through **phagocytosis**. On the other hand the cells that are injured die in an uncontrolled manner called **necrosis**. In necrosis the cell swells and bursts spilling its contents, and elicits an inflammatory response. In most animals, apoptosis is necessary during development and the development is not completed without apoptosis.

Most cells are programmed to undergo apoptosis. It is a highly regulated process and depends on a number of signals inside and outside the cell. Numerous genes are involved in regulating apoptosis. Some of these genes encode enzymes called caspase that cleave proteins at specific sites. The caspases are often produced as procaspase that is activated by another caspase. When one caspase is activated, it cleaves other procaspases that stimulate more caspase activity. This cascade of activity triggers the proteins involved in such cell functions as supporting nuclear membrane and cytoskeleton. The caspases also activate a protein that often keeps the DNAse enzyme in an inactive form. This leads to the breakdown of cellular DNA that leads to cell death.

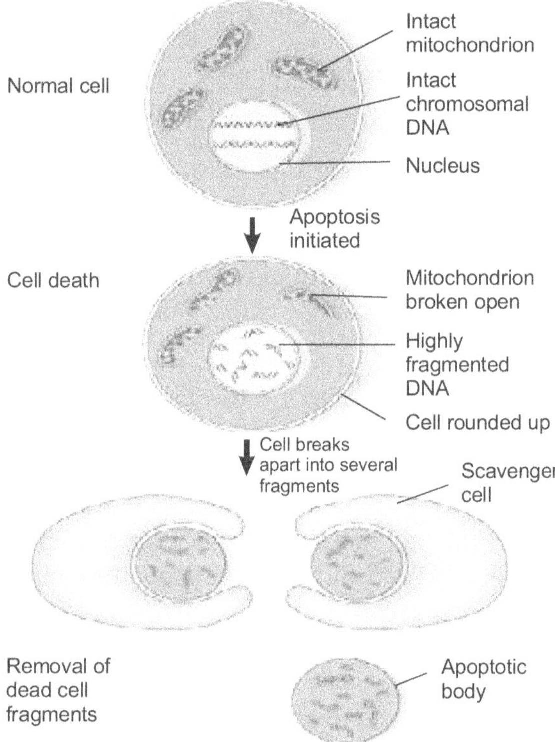

Figure 14.5   The process of apoptosis
(*Source*: www.apoptosisinfo.com)

There are a number of signals that can trigger apoptosis. For example, infection by a virus can activate immune cells to secrete substances which stimulate the cell to undergo apoptosis. Similarly, DNA damage can induce apoptosis and thereby prevent the replication of the mutated sequences. Accumulation of a misfolded protein in the endoplasmic reticulum and damage to mitochondria also can trigger apoptosis. Apoptosis is controlled by cell–cell signalling system. Mutations in genes that regulate apoptosis often stimulate cancer leading to the failure of apoptosis that normally eliminates cancer cells.

Summary

- Developmental process in an organism is a precisely regulated phenomenon.
- It is carried out by the coordinated activities of genes.
- The activities of genes lead to different development patterns, starting from the egg or zygote to the fully developed adult. Initially each cell in the embryo is totipotent.
- All cells of an organism are identical in their genetic content and the development proceeds by differential expression of these genes.
- Cloning experiments demonstrate that in most animals the complete set of genetic information is retained during development and the genetic material is not lost or altered but there is selective expression of genes.
- *Drosophila* is used as a model system to study the genetics of development.
- In *Drosophila,* the entire body plan is established simultaneously at the blastoderm stage.
- The genes involved in the development in fruit flies are egg-polarity genes, and genes determining the dorsal–ventral axis and antero-posterior axis.
- Segmentation genes control the differentiation of the embryo into individual segments. These genes affect the number and organization of the segments.
- The identity of an individual segment is determined by *homeotic* genes.
- Temperature-sensitive mutant studies indicate that temperature-sensitive periods occur during the third instar stage of larval

development. After this period the course of development is not affected by mutations.

✧   In development, the regions of the embryo are determined successively. The genes and gene products are in a cascade of control.

✧   Cell death is initiated by the cell itself and is termed apoptosis.

REVIEW QUESTIONS

1. What experiments suggested that genes are not lost or permanently altered in development?

2. Explain the establishment of dorso-ventral axis in *Drosophila* and the role of dorsal protein.

3. List the three major classes of segmentation genes and outline the functions of each.

4. What role do the homeotic genes play in the development of *Drosophila*?

5. How is apoptosis regulated?

6. What would be the effect of an increase in the number of copies of *bicoid* gene on development?

7. What is the cascade of gene regulation controlling development?

15

BACTERIAL GENETICS

Objectives

- ⚙ To study the recombination process and gene transfer in bacteria
- ⚙ To make use of gene-transfer mechanisms to map their genome

Key Terms

Prototroph	auxotroph	plasmid
episome	F factor	conjugation
transformation	transduction	Hfr transfer
transformants	bacteriophage	virulent
lytic cycle	prophage	lysogenic cycle
sexduction	transducing phage	merozygote
hemizygote	transforming principle	
generalized transduction	specialized transduction	
non-virulent (temperate phage)		

INTRODUCTION

The genetic systems of bacteria and viruses have contributed much to the understanding of molecular genetics. They are important tools for studying the nature of the genes in complex organisms. There are numerous advantages in using bacteria and viruses for genetic studies. Some of them are rapid reproduction; genomes are small and haploid genome allows mutations to be expressed directly; asexual reproduction allows the isolation of pure strains; culturing in the laboratory is simple; they can be genetically engineered.

BACTERIAL CULTURE

Microbiologists have developed several culture media to culture bacteria. Culture media contain a carbon source, essential elements such as nitrogen and phosphorus, certain vitamins and other nutrients and ions. Wild-type bacteria are called prototrophs, since they can grow on minimal medium. This is a medium that contains simple ingredients to synthesize all the compounds required for growth and reproduction. The mutant strains are called auxotrophs that lack one or more enzymes for synthesizing essential molecules. They can grow in a medium which is supplemented with one or more nutrients. For example, an auxotropic strain that is unable to synthesize the amino acid methionine will not grow in a minimal medium but will grow in a medium with methionine. Complete medium contains all the substances required by bacteria for growth and reproduction.

Bacteria can be grown either in a test tube containing sterile liquid medium or in a Petri plate containing the solid agar medium. When grown in Petri plates, it is possible to isolate and count the bacteria. In a process called plating, a dilute solution of bacteria is spread over the surface of agar-filled Petri plate. Each bacterium grows and divides giving rise to visible clumps of genetically identical cells called a colony. Now these pure strains can be isolated by collecting bacteria from single colony and transferring them to new vials or Petri plates.

BACTERIAL GENOME

Bacteria are unicellular organisms that lack a nuclear membrane. In most bacteria, the genome is a circular chromosome that contains a single DNA molecule with several million base pairs. The *E. coli* genome has 4.6 million base pairs approximately.

In addition to a chromosome, many bacteria possess plasmids which are small, circular DNA. Plasmids may be single or many in number. Plasmids play an important role in the growth and life cycle of bacteria. They are used extensively in genetic engineering and also play a role in spreading antibiotic resistance among bacteria. Plasmids possess thousands of base pairs and replicate independently. Plasmid has its own origin of replication, and replication proceeds from the origin in one or two directions until the entire plasmid is copied. Episomes are plasmids that can either freely replicate or integrate into the bacterial chromosome. For example, the F factor (fertility factor) of *E. coli* is an episome.

GENE EXCHANGE AND RECOMBINATION IN BACTERIA

Generally bacteria reproduce by binary fission by amitosis without any exchange of genetic material. In 1946, Joshua Lederberg and Edward Tatum demonstrated that bacteria can exchange genetic material similar to sexual reproduction. Bacteria undergo recombination by three different ways.

1. **Conjugation** is the direct transfer of genetic material from one bacterium to another. The DNA is transferred from a donor bacterium to a recipient and there is no reciprocal exchange of genetic material.

2. DNA in the surrounding medium is taken up by **transformation** and recombination takes place between the introduced genes and the bacterial chromosome.

3. In **transduction,** the bacteriophages (bacterial viruses) transfer genes from one bacterium to another. The introduced DNA undergoes recombination with the bacterial chromosome.

CONJUGATION

Lederberg and Tatum studied strains of *E.coli* possessing auxotropic mutations. The Y10 strain required the amino acids threonine, leucine and the vitamin thiamine but do not require vitamin biotin or the amino acids phenylalanine and cysteine. The genotype of Y10 strain can be written as thr^- leu^- thi^- bio^+ phe^+ cys^+. The Y24 strain required biotin, phenylalanine and cysteine and did not require threonine, leucine and thiamine. The genotype of Y24 can be written as thr^+ leu^+ thi^+ bio^- phe^- cys^-. When Y10 or Y24 were plated on minimal medium separately, they never grew since Y10 required threonine, leucine and thiamine which were absent in the minimal medium. On the other hand, Y24 was unable to grow on minimal medium since it required biotin, phenylalanine and cysteine. But when both the strains were mixed and plated on minimal medium, a few colonies grew in the minimal medium and these prototropic bacteria were of the genotype thr^+ leu^+ thi^+ bio^+ phe^+ cys^+. Lederberg and Tatum concluded that the production of prototrophs is due to recombination and gene transfer between the auxotrophs. This was confirmed by Bernard Davis. He constructed a U-tube that was divided into two compartments by a filter with fine pores. The pore size allowed liquid to pass through and not bacteria. Two auxotrophic strains of bacteria were introduced on opposite sides of the filter and suction was used for the free movement of the medium between the compartments. The bacteria plated out later did not grow in the minimal medium. It was revealed that the exchange of bacterial genes required physical contact. This type of gene exchange requiring cell-to-cell contact is termed conjugation.

F Factor Transfer

In bacteria, conjugation depends on a fertility factor the F element or F factor. This factor is present in donor bacteria that are called F+ cells and the recipient bacteria that lack the F factor are called F− bacteria. The F factor contains an origin of replication and a number of genes required for conjugation. When F+ cell comes in contact with F−, it produces sex pili that makes contact with F− cell. A conjugation tube is formed between them and the gene transfer takes place through it (Figure 15.1). Transfer is initiated when one of the DNA strands is nicked at an origin (*oriT*). One end of the nicked DNA separates from the circle and passes into the recipient cell. Replication takes place on the nicked strand proceeding around the circular plasmid thereby replacing the transferred strand.

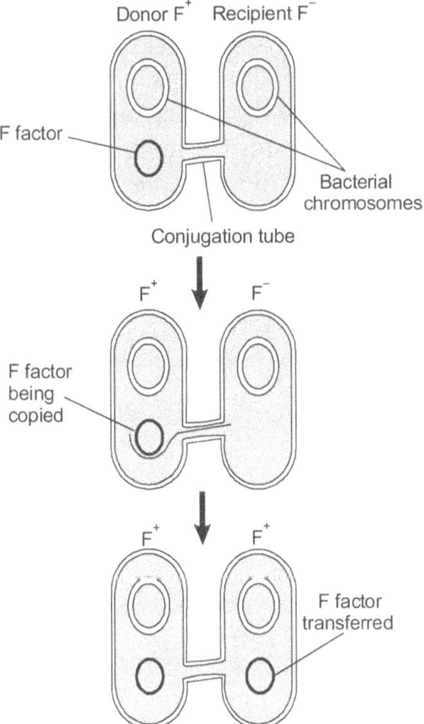

Figure 15.1   Conjugation in bacteria
(*Source*: www.cliffnotes.com)

The *oriT* site always enters first followed by the rest of the strand. So the gene transfer occurs in one particular direction. Inside the recipient cell the single strand duplicates into a double- stranded plasmid. In most cases, the only genes transferred during conjugation between an F^+ and F^- cell are those on the F factor.

Hfr Transfer

In Hfr strains or high-frequency recombination the F factor integrates into the major bacterial chromosome. The Hfr cells behave like F^+ cells and undergo conjugation. In conjugation between Hfr and F^- cells, the integrated F factor is nicked, and the end of the nicked strand moves to the F^- cell. Since the F factor is linked to the bacterial chromosome, the chromosome also enters the F^- cell following the F factor, but how much of the chromosome enters the F^- cell depends on the length of the time of conjugation. Inside the recipient cell, the donor DNA is replicated and crossing over between the original chromosome of the F^- cell and the transferred chromosome takes place. After crossing over, the donated chromosome is degraded and the recombinant recipient's chromosome remains, is replicated and is passed on further (Figure 15.2).

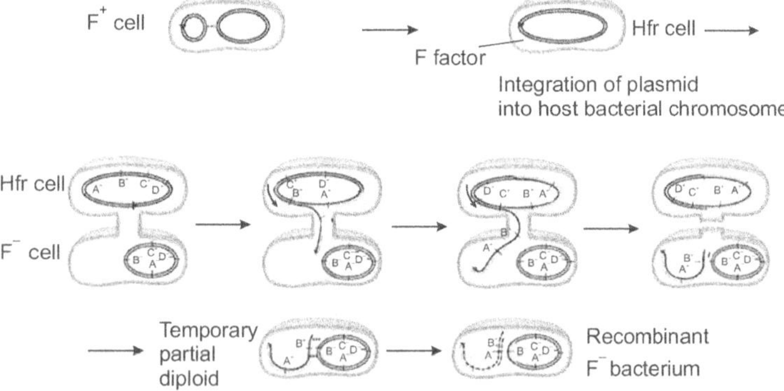

Figure 15.2   Hfr transfer
(*Source*: www.fix.cog.miami.edu)

F' Transfer (Sexduction)

When an F factor does excise from the bacterial chromosome, a small amount of the bacterial chromosome may be removed with it, and these chromosomal genes will be carried with the F plasmid. Cells with F plasmid and some bacterial genes are called F prime (F'). For example, if an F factor integrates into the chromosome adjacent to the chromosome's *lac* operon, the F factor may pick up *lac* genes when it excises, becoming F' *lac*. During conjugation between F' *lac* cell and an F⁻cell, the F plasmid is transferred to the F⁻ cell. This is called sexduction. It produces partial diploids or merozygotes or hemizygotes, which are cells with two copies of some genes, one on the bacterial chromosome and the other on the transferred plasmid.

Interrupted Conjugation and Gene Mapping

The conjugation process involving Hfr and F⁻ cells allows gene mapping. In *E. coli* the transfer of entire chromosome requires about 100 minutes; if conjugation is interrupted before that, only part of the chromosome will pass on to the F⁻ cell. The transfer of chromosome always begins within the integrated F factor and proceeds in one direction. So the genes are transferred according to their sequence on the chromosome. The time required for each gene indicates the position of the gene. The bacterial map is constructed with interrupted conjugation and the basic unit of distance is a minute.

TRANSFORMATION

Transformation was first reported by Frederick Griffith in 1928. While working with *Streptococcus pneumoniae,* he came across two strains of bacteria the virulent S type and a non-virulent R type. The transforming principle from S type could convert the R type to S type. The principle was later identified as the DNA from S type.

Transformation is the uptake of DNA from the medium and its incorporation into the bacterial chromosome or plasmid. The cells that take up DNA are called competent. M. Fox and his co-workers proposed a model to explain transformation. During the process of transformation, the donor DNA binds to the cell surface temporarily (Figure 15.3).

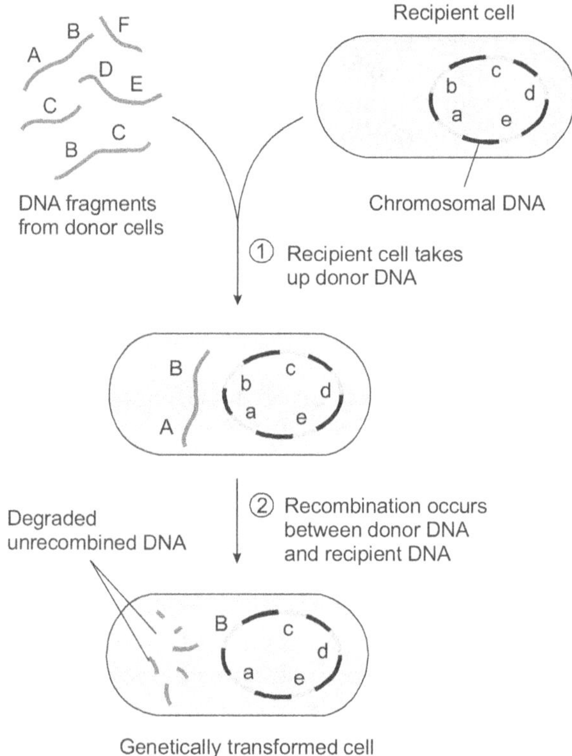

Figure 15.3   Bacterial transformation
(*Source*: www.faculty.ircc.com)

The donor DNA finds the homologous sequence in the host DNA and gets covalently linked to host DNA. This is achieved by the creation of single-strandedness by denaturation in both the DNAs. Alternatively, one strand could be nicked by an endonuclease and then digested by an exonuclease to expose the complementary strand. An additional wedging protein may maintain the single strand in an extended state. The enzyme DNA ligase joins the free ends of the chromosomal segment. The end product of transformation is a duplex in which one strand is derived from donor DNA and the other from host DNA.

Mapping by Transformation

Transformation is also used to map bacterial genes. It requires two strains of bacteria differing in genetic traits. Let us suppose the donor strain is an auxotroph designated as $a^- b^- c^-$ and the recipient strain is a prototroph with $a^+ b^+ c^+$. The donor DNA fragments are added to the medium. The donor DNA enters the recipient and undergoes recombination with the recipient's (transformants) chromosome. Genes are mapped by noting the rate at which two or more genes are transferred together (co-transformed). When a DNA is fragmenting, it is more likely that genes which are closer remain in the same fragment and get transferred together. Genes that are farther apart may not be in the same fragment and so cannot be transferred together. The frequency of co-transformation can be used to map bacterial genes. Genes a and b, and b and c are often co-transformed and a and c are not. Then the order of the gene is a,b,c.

TRANSDUCTION

Transduction involves the transfer of gene from one bacterium to another through a virus usually called a bacteriophage. When a bacteriophage infects a bacterium it can act as a virulent form and undergo lytic cycle or it can act as a non-virulent form (temperate phage) and integrate into the bacterial chromosome. This is called a prophage and undergoes lysogenic cycle. When a prophage leaves a bacterium it takes along with it the host DNA (transducing phage) and transfer the bacterial DNA to another bacterium when it infects the new bacterium. Recombination occurs and the recombinant bacterium is a transductant. This process is called transduction (Figure 15.4). Transduction may be either generalized transduction or specialized transduction.

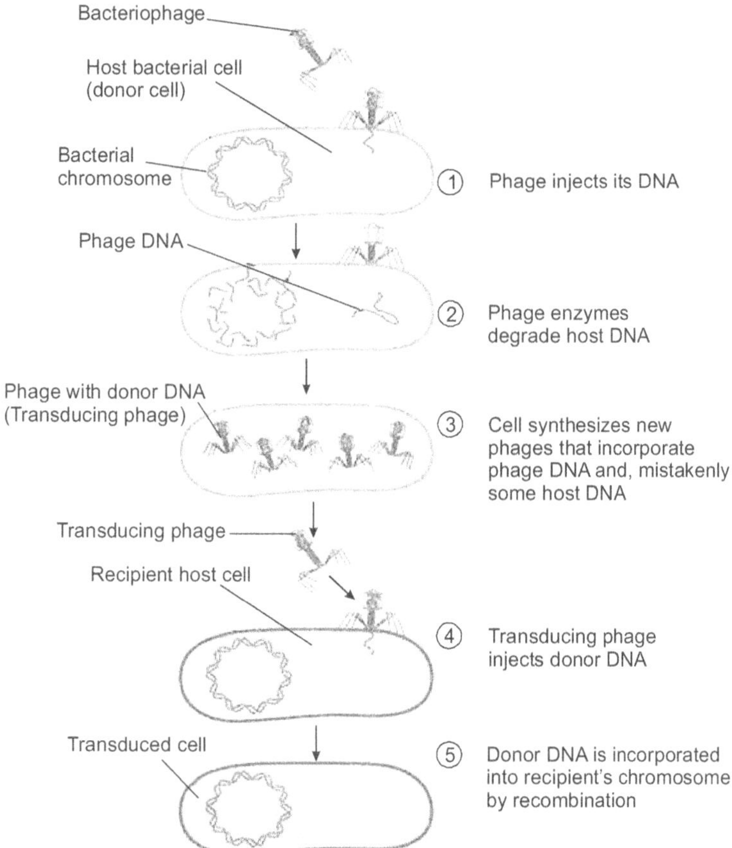

Figure 15.4   Transduction in bacteria
(*Source*: www.faculty.ircc.com)

In specialized transduction only few genes are specifically transferred. In generalized transduction, the genes of the donor bacterium replace the recipient genes in the recipient chromosome. Recombinants will be formed when the inserted DNA carries donor DNA that differ from recipient genes. The phenomenon is called generalized transduction because random pieces of host DNA have been transduced.

In specialized transduction the genes near prophage sites on the bacterial chromosome are transferred. When the prophage deintegrates, it may excise from the bacterial chromosome, carrying with it a small part of the bacterial

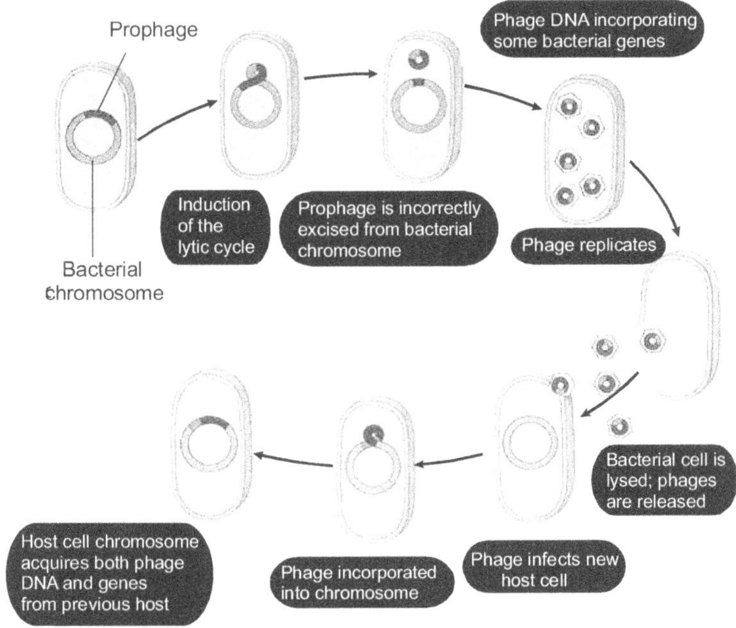

Figure 15.5   Specialized transduction
(*Source*: www.faculty.ircc.com)

DNA adjacent to the site of prophage integration. Then it injects it into another bacterial cell in the next cycle of infection. For example, the lambda phage (λ) integrates into the *E. coli* chromosome at the attachment site (*att*). During excision, the genes on either sides of *att* may also be excised along with the phage chromosome (Figure 15.5). In *E.coli*, these genes are usually the *gal* and *bio* genes and these genes also will be transferred along with the phage DNA.

Summary

- Bacteria are well suited for genetic studies.
- They are small, have a haploid genome and multiply rapidly.
- Individual bacteria can be spread on Petri plate and colonies can be grown.
- The bacterial genome consists of a single, circular and double-stranded DNA.
- Plasmids are small pieces of DNA that replicate independently of the major chromosome.
- Bacteria transfer genes by means of conjugation, transformation and transduction.
- Conjugation is the union and transfer of genetic material between two bacterial cells.
- F^+ cells are donors and F^- cells are recipients. An Hfr cell has F factor integrated into the genome.
- An F´ cell has F factor that has excised from the bacterial genome and carries some bacterial genes.
- The rate at which the genes are transferred determines the gene sequence, and gene mapping can be carried out by interrupted mating.
- In transformation, bacteria take up DNA from the environment.
- In transduction genes are transferred from one bacterium to another through a phage.
- In generalized transduction, bacterial genes become incorporated into phage coat and are transferred to other bacteria.
- In specialized transduction, DNA near the site of phage integration on the bacterial chromosome is transferred from one bacterium to another.

REVIEW QUESTIONS

1. State the differences between F⁺, F⁻, Hfr and F′ cells.

2. By what mechanism are F⁻ cells converted to F⁺ cells and F⁺ cells to Hfr cells?

3. Compare and list the similarities and differences between conjugation, transformation, transduction and sexduction.

4. Give the basic differences between generalized transduction and specialized transduction.

5. Explain how interrupted conjugation and transformation can be used to map bacterial genes.

6. Differentiate lytic cycle from lysogenic cycle.

7. Sketch F⁺ transfer and Hfr transfer.

16

CYTOPLASMIC INHERITANCE AND MATERNAL EFFECTS

Objectives

- ✿ To study the non-Mendelian inheritance
- ✿ To analyse the types of cytoplasmic inheritance
- ✿ To observe the patterns of maternal effect

Key Terms

kappa particles	killer	sensitive
paramaecin	sigma	mtDNA
maternal effect	dextral	sinistral
tumorous head	cytoplasmic	
non-Mendelian inheritance	extranuclear inheritance	
plasma gene/plasmon		

INTRODUCTION

Nuclear DNA is the most important and very nearly the universal genetic material. Mendel's principles predict the types of offspring that will be produced in a genetic cross. However, not all the genetic material of a cell is found in the nucleus; some traits are encoded by genes present in the cytoplasm. These traits are transmitted by cytoplasmic inheritance. Extrachromosomal inheritance or cytoplasmic inheritance is defined as non-Mendelian inheritance, usually involving DNA in cytoplasmic organelles. A few bacteria and viruses are also agents for extranuclear inheritance.

Carl Correns first reported cytoplasmic inheritance in 1909 while studying the inheritance of leaf variegation in the four-o'clock plant *Mirabilis jalapa*. In 1950, Sanger and his colleagues also suggested the possible role of cytoplasm in inheritance. For example in *Chlamydomonas,* inheritance of certain characters is controlled by non-chromosomal genes. In some animals and plants certain traits are inherited independent of the nuclear genes. In such cases the cytoplasm seems to contain DNA. These may be mitochondria, plastids or foreign organisms. The cytoplasmic hereditary particles are generally termed as plasmon and the hereditary units in particular are called the plasma genes.

FEATURES OF CYTOPLASMIC INHERITANCE

Cytoplasmic inheritance differs from nuclear inheritance in several important respects.

- ✿ A zygote inherits nuclear genes from both parents. Sperm usually contributes only a set of nuclear genes from the male parent. In most organisms, all the cytoplasm is inherited from the mother. So cytoplasmically inherited traits are present in both males and females and are passed from mother to offspring and never from father to offspring.

- ✿ Reciprocal crosses give different results for cytoplasmic inheritance. The reciprocal cross ratios deviate from the Mendelian inheritance pattern.

- ✿ Cytoplasmic genes are not evenly distributed in cell division. So there will be extensive phenotypic variation. The cells of an individual will contain the cytoplasmic genes in various proportions. The female has more influence on the traits as it carries more cytoplasm and cytoplasmic organelles than the male.

✡ Chromosomal genes occupy particular loci and exhibit linkage. Extranuclear genes fail to show linkage.

MITOCHONDRIA AND CYTOPLASMIC INHERITANCE

Organelles like mitochondria and chloroplasts contain DNA. Mitochondria are presently considered as living organisms. They are usually small cytoplasmic organelles with cristae arising from the inner mitochondrial membrane. They are about the same size as bacteria and occur in cells of eukaryotes but not in bacteria and viruses. They contain a small amount of unique DNA that has remained autonomous outside the nuclear genome.

The mitochondrial genome is small and codes only for a limited number of structures and functions. Mitochondria contain a distinctive protein-synthesizing apparatus with specific ribosomes, tRNAs, and aminoacyl-tRNA synthetases; it exhibits antibiotic sensitivity like that of bacteria. The mitochondrial DNA (mtDNA) is a circular molecule. The mtDNA range in size about 16 kb in mammals up to several hundred kilobase pairs in higher plants. The mtDNAs of humans, mice and cattle exhibit the same basic organization of genetic information. Each mtDNA has 2 rRNA genes, 22 tRNA genes and 13 structural genes. The entire mtDNA is equivalent to one operon in bacteria. There are a number of proteins in mitochondria that are encoded by nuclear genes.

The human mitochondrion contains about 15,000 nucleotides of DNA and encodes 37 genes. On the other hand, the nuclear DNA contains 3 billion nucleotides and encodes 35,000 genes. The amount of mitochondrial DNA (mtDNA) is very small.

Kappa Particles in Paramecia

Paramecia are favoured organisms for genetic investigation. They reproduce both by sexual and asexual means. In the sexual phase, paramecia conjugate periodically and transfer genetic material from one cell to another. Paramecia have two kinds of nuclei: a large vegetative macronucleus and a small micronucleus which undergoes meiosis. It is possible to make sexual crosses through which nuclear DNA is transferred from a donor to a recipient, resulting in heterozygous progeny, that is, AA × aa → Aa. A process of self-fertilization, called autogamy, results in the complete homozygosis of the resulting progeny. Following meiosis, the cells are haploid, but through autogamy they become homozygous diploids. This provides the basis for comparing extranuclear and nuclear inheritance.

T.M. Sonnebonn and others have investigated a persistent extranuclear effect in Paramecium. Some strains of *Paramecium aurelia* produce a substance that has a lethal effect on members of other strains of the same species. Paramecia that are capable of producing the toxic substance are called killers. The toxic effect decreases by repeated cell division. Killers possess particles called kappa. These are shown to be symbiotic bacteria, have been named *Caedobacter taeniaspiralis* (the killer bacterium with the spiral ribbon).

A killer bacteria produces a "toxic substance" namely paramecin. When a "sensitive" strain of Paramecium is exposed to the medium where killer strains were kept, the sensitives were killed. Paramecin, which has no effect on killers, is associated with kappa particles. The kappa bacteria possess a refractile protein containing 'R' body and are called "brights", because they are infected with a virus that controls the synthesis of viral protein. The virus is toxic to sensitive paramecia but is not toxic in "non-bright" bacteria.

The kappa particles multiply only in organisms that carry the dominant nuclear allele K. When killers conjugate with sensitives for short time, no cytoplasmic exchange occurs. Two kinds of clones emerge: one from the original killer cell, which contains allele K (KK) and kappa, and the other from the original sensitive cell, which carries the allele k (kk) and lacks kappa. Following autogamy, half the progeny of the killers are killers and half are sensitive paramecia. All progeny of sensitives are sensitives. Since no cytoplasm was transferred in this conjugation, only the cells from original killers inherit kappa particles. Kappa cannot reproduce in cells unless a K allele is present in the nucleus (Figure 16.1).

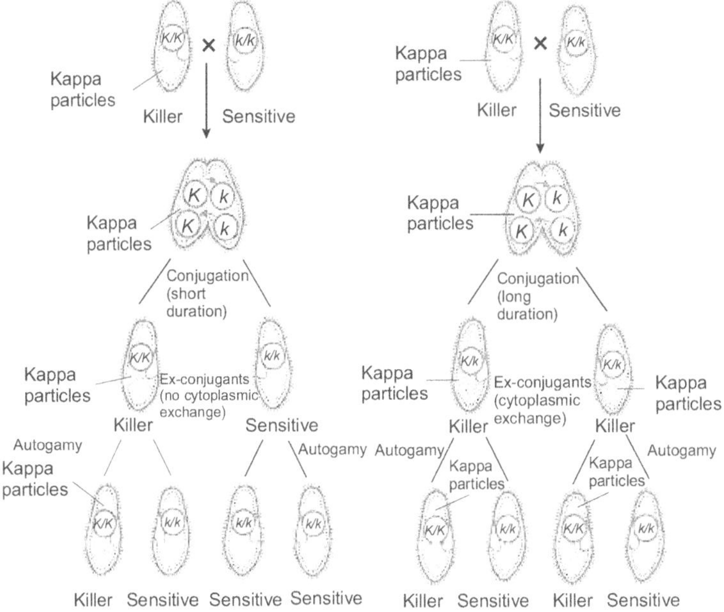

Figure 16.1   Inheritance of kappa particles in *Paramecium* (*Source*: Adapted from Burns, 1969)

Under some conditions, conjugation persists much longer; cytoplasm as well as nuclear genes is exchanged between conjugates. When the conjugants are KK and kk, alleles K and k are exchanged and both ex-conjugants are Kk. Cytoplasmic exchange has transferred kappa from the killer to the non-killer cell. A prolonged conjugation results in non-killer also to receive kappa. But later autogamy further produces homozygous KK and kk cells, which produce clones of killers or non-killers, respectively.

Sigma Factor in *Drosophila*

Most fruit flies can be exposed to carbon dioxide for long hours, without injury. However L'Heritier and Teisser discovered a true breeding strain of *Drosophila* which was sensitive to carbon dioxide. The sensitive flies, when exposed to carbon dioxide for a short period, become unconscious in a characteristic way, with their legs becoming paralysed. When reciprocal crosses were made between carbon dioxide-sensitive and normal strains, it could be shown that the trait was inherited only from female parent. Sensitive mothers always give sensitive progeny. It was shown that the sensitivity can be contributed to a viruslike particle called sigma found in the cytoplasm of the cells of a sensitive fly.

The transmission of the sigma factor is through egg cytoplasm and its reproduction depends on initial supply and on suitable temperature of 20°C, because it is heat-labile at high temperature. It does not need a specific gene and may be found associated with different genotypes. A sensitive fly retains its sensitive trait, even when all its chromosomes are replaced by those of normal fly, suggesting that sigma factor is located in the cytoplasm and has the property of plasma genes. The sigma factor is 0.07 micron in diameter and contains DNA responsible for heredity.

Yeast

In yeast cells, 10–20 per cent of the cellular DNA is localized in a single mitochondrion. Mitochondrial DNA has properties different from those of nuclear DNA. The life cycle of normal baker's yeast, *Saccharomyces cerevisiae*, includes a haploid and diploid phase. Mating normally occurs between vegetative haploid cells of opposite mating type (Aora). These cells fuse to form vegetative diploid cells that divide by mitosis. Cell division is usually unequal, with a small "daughter" cell budding from a larger "mother". Both cells, however, are identical in nuclear composition.

The first mutant found in yeast, a small colony type called petite, has now provided the best evidence known, for mitochondrial mutations. Petites are defective in their ability to utilize oxygen in the metabolism of carbohydrates. For example, when glucose is in the medium, petite yeast will grow to only small-sized colonies. Enzyme analysis indicates that mitochondria lack the enzyme cytochrome oxidase associated with mitochondria. Petite strains when analysed show only a small proportion of GC base pairs and a preponderance of repetitive AT base pairs. Absence of cytochrome oxidase indicates that mutational changes in mitochondrial DNA lead to heritable alterations in mitochondrial phenotypes.

Mutations other than those causing petites can be induced in yeasts and transmitted by the cytoplasm. For example, resistance to the antibiotics chloramphenicol and erythromycin has been induced.

Yeast chromosomal genes specify most enzymes of mitochondria. Petite yeast strains with damaged DNA continue to synthesize abnormal mitochondrial DNA. This indicates that the proteins needed for mitochondrial DNA replication are not encoded by mitochondrial DNA. Similarly, petite strain continues to synthesize the Kreb's cycle enzymes that are located in the mitochondria. So the control must come from chromosomal genes.

The circular mtDNA of the yeast *Saccharomyces cerevisiae* is over five times larger than that of mammals. The organization of yeast mtDNA is similar to that of mammalian mtDNA. Two yeast mtDNA genes encoding cytochrome *b* and subunit 1 of cytochrome oxidase are very large and contain several long intron sequences. The large yeast mitochondrial genome also encodes more proteins than the mammalian mtDNAs.

Ephruassi and his colleagues conducted experiments to show that nuclear genes behaved according to a Mendelian pattern of inheritance but the petite phenotype yielded a ratio of almost 0 wild type : 4 petite, a non-Mendelian segregation.

MATERNAL EFFECTS

A genetic phenomenon that is sometimes confused with cytoplasmic inheritance is maternal effect, in which the phenotype of the offspring is determined by the genotype of the mother. On the other hand in cytoplasmic inheritance, the genes for a trait are inherited from both parents, but the offspring's phenotype is determined not by its own genotype but by the genotype of the mother. Eggs and embryos are expected to be influenced by the maternal environment in which they develop. Certain potentialities of the egg are known to be determined before fertilization, and in some cases these have been influenced by the surrounding maternal environment. Such predetermination by genes of the mother, rather than those of the progeny is called maternal effect. If a maternal effect is involved, results from reciprocal crosses will be different from each other, with genes of the mother being expressed.

Shell Coiling in Snail

One of the best examples of maternal effect is that of the direction of coiling in shells of the snail *Limnaea peregra*. Some strains of this species have dextral shells, which coil to the right; others have sinistral shells, which coil to the left. This characteristic is determined by the genotype of the mother (not her phenotype) rather than by the genes of the developing snail. Allele S^+ for right-handed coiling is dominant over allele S for coiling to the left.

When crosses were made between females coiled to the right and males coiled left, the F_1 snails were all coiled to the right. The usual 3 : 1 ratio was not obtained in the F_2 because the phenotype SS was not expressed. Instead, the pattern determined by the mother's (P_1) genes (S^+S^+) was expressed in the F_1, and the F_1 mother's genotype (S^+S) was expressed in the F_2.

When individuals were inbred, only progeny that coiled to the left were produced. When the S^+S^+ or S^+S snails were inbred, however, they produced offspring that all coiled to the right. From the reciprocal crosses between left-coiling females and right-coiling males, all F_1 progeny were coiled to the left. The F_2 snails coiled to the right; but, when each F_2 snail was inbred, those with the genotype SS produced progeny that coiled to the left (Figure 16.2).

Further investigation of coiling in snails has shown that the spindle formed in the metaphase of the first cleavage division influences the direction of coiling. The spindle of potential "dextral" snails is tipped to the right, but that of "sinistral" snails is tipped to the left. This difference in the arrangement of the spindle is controlled by the genes of the mother. They determine the orientation of the spindle, which in turn influences further cell division and results in the adult pattern of coiling. The actual phenotypic characteristic, therefore, is influenced directly by the mother, with no immediate relation to the genes in the egg, sperm or progeny.

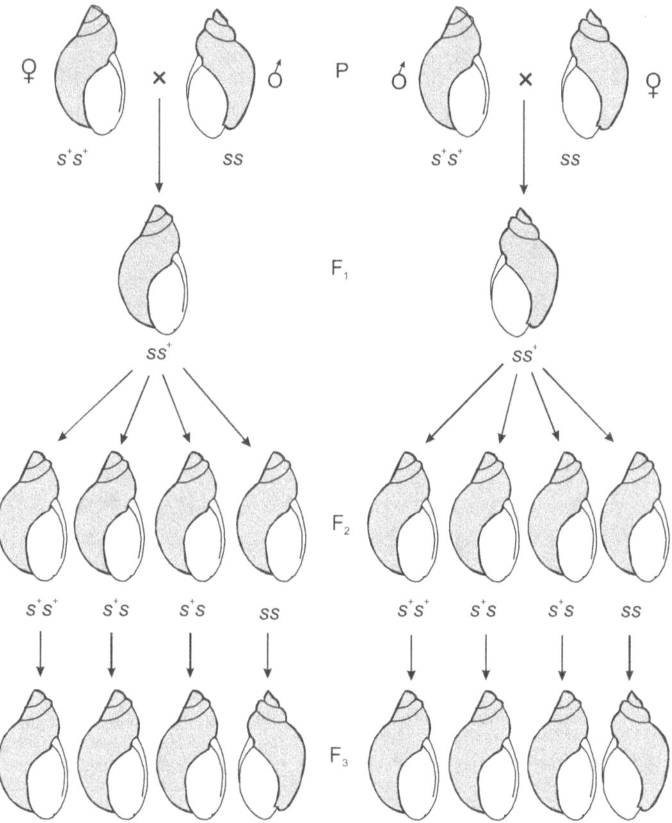

Figure 16.2   Shell coiling in snails
(*Source*: www.nitro.biosci.arizona.edu)

Maternal Effect in *Drosophila*

In *Drosophila melanogaster,* an abnormal growth in the head region was observed. This trait is called "tumorous head" (Tu-h). When reciprocal crosses were conducted, a maternal effect was indicated. Tu-h females mated separately with three wild type males and about 14 to 52% abnormal flies were obtained in the first generation. From the reciprocal cross between Tu-h males and the same three wild type females, about 0–1% of tumorous head flies were obtained. Further studies demonstrated maternal effect. Genes of the mother were exerting an influence in the direction of abnormal growths on the heads of adult progeny during the first 22 hours of development. Two major genes were found to control the tumorous head trait: (1) a sex-linked gene at 64.5 map units on the X chromosome controlling the maternal effect and (2) a structural gene at 58 map units on the third chromosome controlling the tumorous head phenotype.

Summary

- Extrachromosomal inheritance or cytoplasmic inheritance is defined as non-Mendelian inheritance, usually involving DNA in cytoplasmic organelles.

- A few bacteria and viruses are also agents for extranuclear inheritance.

- The cytoplasmic hereditary particles are generally termed as plasmon and the hereditary units in particular are called the plasma genes.

- Cytoplasmic inherited traits are present in both males and females and are passed from mother to offspring and never from father to offspring.

- The reciprocal cross ratios deviate from the Mendelian inheritance pattern.

- Organelles like mitochondria and chloroplasts contain DNA and are involved in cytoplasmic inheritance.

- In *Paramecium,* cytoplasmic exchange has transferred kappa particles from the killer to the non-killer cell.

- In *Drosophila* the sensitivity to carbon dioxide was inherited only from female parent. Sensitive mothers always give sensitive progeny.

- The first mutant found in yeast, a small colony type called petite, has provided the best evidence for mitochondrial inheritance.

- ✿ A genetic phenomenon that is sometimes confused with cytoplasmic inheritance is maternal effect, in which the phenotype of the offspring is determined by the genotype of the mother.

- ✿ One of the best examples of a maternal effect is the direction of coiling in shells of the snail *Limnaea peregra*. This characteristic is determined by the genotype of the mother rather than by the genes of the developing snail.

- ✿ In *Drosophila melanogaster*, the inheritance of abnormal tumorous head (Tu-h) indicates a maternal effect.

REVIEW QUESTIONS

1. Differentiate cytoplasmic inheritance from Mendelian inheritance.

2. List the unique features of extranuclear inheritance.

3. Why are mitochondria considered to be hereditary units?

4. Show the inheritance of kappa particles in *Paramecium*.

5. What is the sigma factor in fruit fly?

6. State the genetic cause for the occurrence of petite yeasts.

7. Distinguish between maternal effect and cytoplasmic inheritance.

8. Explain the maternal inheritance in shell coiling in snail.

17

POPULATION GENETICS

Objectives

- ✿ To learn the assumptions and nature of Hardy–Weinberg equilibrium
- ✿ To calculate the allelic frequencies and test whether a population is in Hardy–Weinberg equilibrium
- ✿ To analyse the factors that alter Hardy–Weinberg equilibrium

Key Terms

gene pool	gene frequencies	hemizygous
mutation	selection	migration
random drift	panmictic	fitness
gametic selection	zygotic selection	selection coefficient
adaptive value	selective value	mutational equilibrium
founder effect	random genetic drift	
genotypic frequencies	Hardy–Weinberg equilibrium	

INTRODUCTION

The inheritance of individual genes is governed by Mendelian principles, but the frequencies of these genes in a population may be influenced by many factors like the size of population and frequency of a particular gene and several other factors. The distribution of a particular gene (its alleles) in time and space is not dependent on individuals carrying this gene but is governed by the properties of the population consisting of individuals carrying this gene, i.e., its alleles.

A population consists of a community of sexually interbreeding organisms inhabiting a geographical region. This population carries particular genes. These populations were called "Mendelian populations" by Sewall Wright, because individuals belonging to these populations follow Mendelian principles of inheritance.

BEHAVIOUR OF GENES IN A POPULATION

Gene pool and gene frequencies are two important attributes of a population. A gene pool is the sum total of genes in gametes of a population. The gene pool is transferred from one generation to the other through the sample drawn from a gametic pool. Gene pool consists of a large number of genes which will vary in their frequencies. Gene frequencies are defined as proportions of different alleles of a gene in a population, and in a particular generation, these frequencies will depend upon their frequencies in the preceding generation. These frequencies also depend on the proportion of various genotypes in the total population.

For example, in human MN blood groups, in a sample population of 100 individuals with 50 MM, 20 MN and 30 NN, frequencies of 'M' and 'N' can be calculated. Since each individual will have two homologous chromosomes, each carrying a particular allele, the frequency of 'M' can be calculated by doubling the number of homozygous 'M' blood group type and adding to it the frequency of heterozygous 'MN' blood group type. In this manner the frequency of 'M' will be (50 × 2) + 20 = 120. Similarly the frequency of 'N' will be (30 × 2) + 20 = 80. The relative frequencies of 'M' allele can be worked as $\dfrac{M}{M+N}$ and that of 'N' can be worked out as $\dfrac{N}{M+N}$. Therefore the frequency of 'M' is 120/200 = 0.6 and that of 'N' is 80/200 = 0.4.

The frequencies of genes can also be worked out with the help of the following formula:

Frequency of gene = Frequency of homozygote for that gene + 1/2 frequency of heterozygotes

So,

Frequency of M = 0.5 MM + 1/2 (0.2 MN) = 0.6

Frequency of N = 0.3 NN + 1/2 (0.2 MN) = 0.4

HARDY–WEINBERG EQUILIBRIUM

In 1908, G.H. Hardy, an English mathematician and W. Weinberg, a German physician independently discovered that an equilibrium is established between frequencies of alleles in a random mating population. It was also shown that the relative gene frequencies remained unaltered from one generation to the next, regardless of their dominance and recessive relationship. This equilibrium is explained by Hardy–Weinberg law. It states that "in a large random mating population, the gene frequency and the genotype frequency remain constant generation after generation when there is no mutation, no selection, no migration and no genetic drift." When Hardy–Weinberg law operates, there is no evolution noticed, when the frequencies are altered, there is evolution operating in the population.

Frequencies of Two Alleles of a Single Locus

When two individuals heterozygous for the same gene Aa are crossed (Aa × Aa), they segregate in a ratio 1 AA : 2 Aa : 1 aa. This ratio can be obtained by simple expansion of binomial theorem $(A + a)^2 = 1$ AA : 2 Aa : 1 aa. In this case, frequencies of A and a are the same, i.e., 0.5. The frequencies of 'A' and 'a' are designated as '*p*' and '*q*'. The probabilities of individuals obtained from random mating from such population can be obtained as shown in Table 17.1.

It can be seen from the table that the probability of homozygous dominant (AA) obtained through random mating would be p^2, that of heterozygous (Aa) individuals would be 2pq and that of homozygous recessive (aa) would be q^2. This leads to the formulation of the Hardy–Weinberg equation: p^2 (AA) + $2pq$ (Aa) + q^2 (aa). This equation would be obtained in every case after one generation of random mating, if factors like mutation, selection and migration do not operate. In a particular heterogeneous population where for a particular character complete dominance is expressed, no distinction can be made between homozygous dominant (AA) and heterozygous (Aa) individuals. Therefore, '*p*' cannot be directly determined; instead it can be indirectly determined by first determining the value of '*q*', through simple square root of frequency of the individuals having recessive phenotype. Since $p + q = 1$, p will be equal to $1 - q$. Substituting the value $1 - q$ for p in the Hardy–Weinberg equation, it can be written as $(1 - q)^2 + 2q (1 - q) + 1q^2 = 1$. It is possible to prove algebraically that a population having genotypic frequencies expressed is $p^2 + 2pq + q^2$ will be in equilibrium.

Table 17.1   Frequencies of three genotypes in a population having allele A and a with frequency *p* and *q* respectively

Sperms / Eggs	A (*p*)	a (*q*)
A (*p*)	AA(p^2)	Aa(*pq*)
a (*q*)	Aa(*pq*)	aa(q^2)

Calculation of Gene Frequencies

Distribution of alleles in succeeding generations resulting due to random mating, can be calculated by using Hardy–Weinberg equation $(p^2 + 2pq + q^2)$. PTC tasting can be taken as an example. The chemical phenylthiocarbamide (PTC) is tasted bitter by some persons. This trait is due to a single gene with two alleles, T and t. Taste trait is controlled by the dominant gene T and the non-taster trait by t, the recessive gene. Since T is dominant over t, the tasters can have two genotypes TT and Tt, whereas non-tasters will have only one genotype 'tt'.

Suppose in an initial population of human beings, different genotypes are represented by the ratio 0.40TT : 0.40Tt : 0.20tt. The gene frequencies will be

T = 0.40 (TT) + 1/2 × 0.40 (Tt) = 0.60

t = 0.20 (tt) + 1/2 × 0.40 Tt = 0.40.

The frequencies of T = 0.60 and t = 0.40 will give a genotypic ratio of 0.36 (TT) : 0.48 (2Tt) : 0.16 (tt) in the next generation. This is illustrated in Table 17.2.

Table. 17.2 Frequencies of genotypes of taster (T) and non-taster (t) under conditions of random mating

	p(T) 0.6	q(t) 0.4
p(T) 0.6	p^2 0.36 (TT)	pq 0.24 (Tt)
q(t) 0.4	pq 0.24 (Tt)	q^2 0.16 (tt)

This has been explained in detail in Table 17.3, where frequencies of different mating types and frequencies of different genotypes in the progeny from each mating type are given. This ratio is obtained when there is random mating. It is visible in the table that although the genotype frequencies have changed, the gene frequencies did not change. This demonstrates that after one generation of random mating, equilibrium in the genotypic ratio can be achieved and this will follow the Hardy–Weinberg equation.

Table 17.3 Relative frequencies of the different kinds of offspring produced by the matings in a population with 0.40 TT, 0.40 Tt and 0.20 tt genotypes

Parents		Offspring ratio			Offspring frequencies		
Type of mating	**Frequency of mating**	**TT**	**Tt**	**tt**	**TT**	**Tt**	**tt**
TT × TT	0.16	all 0.16					
TT × Tt	0.32	1/2 (0.16) + 1/2 (0.16)			0.16	0.16	
TT × tt	0.16	all (0.16)				0.16	
Tt × Tt	0.16	1/4 (0.04) + 1/2 (0.08) + 1/4 (0.04)			0.04	0.08	0.04
Tt × tt	0.16	1/2 (0.08) + 1/2 (0.08)				0.08	0.08
tt × tt	0.04	all (0.04)					0.04
	Total				0.36	0.48	0.16

Frequencies of More Than Two Alleles (Multiple Alleles) at a Single Locus

In the above examples, only two alleles at a single locus are considered. The situation becomes complicated when genes with three or more alleles are taken into consideration.

Each allelic frequency should be considered as an element of multinomial expansion. For instance if there are three alleles, A_1, A_2 and A_3 at the locus, with their corresponding gene frequencies p, q and r, respectively, then $p + q + r = 1$. In such a case the trinomial expansion, i.e., $(p + q + r)^2 = p^2 (A_1A_1) + 2pq (A_1A_2) + 2pr (A_1A_3) + q^2 (A_2A_2) + 2qr (A_2A_3) + r^2 (A_3 A_3)$ will represent the genotypic frequencies at the equilibrium stage. This is shown in Table 17.4.

Table 17.4 Frequencies of genotypes when three alleles A_1, A_2 and A_3 are involved with their frequencies being $p = 0.2$, $q = 0.5$ and $r = 0.3$

Gene frequency	$p(A_1)$ 0.2	$q(A_2)$ 0.5	$r(A_3)$ 0.3
$p(A_1)$ 0.2	$p^2 = 0.04$ (A_1A_1)	$pq = 0.10$ (A_1A_2)	$pr = 0.6$ (A_1A_3)
$q(A_2)$ 0.5	$pq = 0.1$ (A_1A_2)	$q^2 = 0.25$ (A_2A_2)	$qr = 0.15$ (A_2A_3)
$r(A_3)$ 0.3	$pr = 0.06$ (A_1A_3)	$qr = 0.15$ (A_2A_3)	$r^2 = 0.09$ (A_3A_3)

Frequencies for X-linked Alleles

For an X-linked locus with two alleles $X^A X^a$, there are five possible genotypes: $X^A X^A$, $X^A X^a$, $X^a X^a$, $X^A Y$ and $X^a Y$. Females possess two X-linked alleles and the expected proportions of the female genotypes can be calculated by using the square of allelic frequencies. If the frequencies of X^A and X^a are p and q respectively, then the equilibrium frequencies of the female genotypes are $(p + q)^2 = p^2$ (frequency of $X^A X^A$) + $2pq$ (frequency of $X^A X^a$) + q^2 (frequency of $X^a X^a$). Males have only one X-linked allele (hemizygous), and so the frequencies of the male genotypes are p (frequency of $X^A Y$) and q (frequency of $X^a Y$). Therefore p^2 is the expected proportion of females with genotype $X^A X^A$, if females make up 50% of the population, then the expected proportion of this genotype in the entire population is $0.5 \times p^2$.

The frequency of an X-linked recessive trait among males is q, whereas the frequency among females is q^2. When an X-linked allele is uncommon, the trait will therefore be much more frequent in males than in females. Consider haemophilia, a blood clotting disorder caused by a recessive allele with a frequency (q) of approximately 1 in 10,000 or 0.0001. At Hardy–Weinberg equilibrium, this frequency will also be the frequency of the disease among males. The frequency of the disease among females, however, will be $q^2 = (0.0001)^2 = 0.00000001$, which is only 1 in 10 million. Haemophilia is 1000 times more frequent in males than in females.

FACTORS ALTERING GENE FREQUENCIES

Gene frequencies are conserved from one generation to the other under certain conditions. Under these conditions frequencies of genotypes reach equilibrium after a single generation of random mating. The conditions include absence of mutation, selection, migration and random drift. This ideal condition never exists and in a large random (panmictic) mating population, changes in gene frequencies do occur. This change can be either directional as in the case of mutation, selection or migration or non-directional as in the case of random drifts. The directional change means a change of gene frequencies progressively from one value to another in either direction. If this change proceeds unchecked, then fixation of one allele and elimination of the other allele happens. The non-directional change cannot predict the fate of an allele from one generation to the other.

MUTATION

Mutation introduces new genes leading to genetic differences in the population. These new genes introduced may or may not persist in the population upon their utility. The gene frequencies will also depend upon this factor. This can be illustrated by the following hypothetical example. If a dominant gene 'A' mutates to 'a', then frequency of 'a' will replace 'A'. Quantitatively, let p_o be the initial frequency of 'A' and 'u' be the mutation rate with which 'A' changes to 'a'. In such a case, 'a' will appear with a frequency of $u \times p_o$ in the first generation. The frequency of 'A' will therefore be reduced by a factor $p_o u$ and become

$$p_o - p_o u = p_o (1 - u)$$

In the next generation there will, therefore, be further change due to the change of 'A' to 'a', thus further reducing the frequency of 'A' by a factor

$$p_o(1 - u) - p_o(1 - u) \times u = p_o(1 - u)(1 - u) = p_o(1 - u)^2$$

In this manner, in 'n' generations, the frequency of 'A' will be reduced to $p_o(1 - u)^n$. Eventually the term $(1 - u)^n$ will approach zero so that 'A' will disappear after several generations, if no reverse mutation takes place and the mutant allele experiences no selection pressure against it. However, if the reverse mutation also takes place with a frequency of 'v' and the initial frequency of 'A' and 'a' are p_o and q_o respectively, in one generation the frequency of 'A' will become

$$p_o + vq_o - up_o$$

and that of 'a' will become

$$q_o + up_o - vq_o.$$

It is obvious that 'a' gains a fraction (up_o) and loses a fraction (vq_o) at the same time. Similarly, 'A' gains a fraction (vq_o) and loses a fraction (up_o). Let us now consider the fate of the frequency of 'a' in the following generations. Let the change in the frequency of 'a' be represented as $\Delta q = up_o - vq_o$.

If p_o is relatively larger than q_o, and u is relatively larger than v, Δq would be fairly high and the frequency of 'a', i.e., q could increase rapidly. This will lead to a situation, where q becomes larger than p so that the value of vq will increase and that of up will decrease. As a consequence, q would diminish gradually and at a certain point, mutational equilibrium will be reached where Δq would become zero. The mutation rates for most genes are low; so change in allelic frequency due to mutation rate in one generation is very small, and long periods are required for a population to reach mutational equilibrium.

SELECTION

The gene frequencies may change due to selection in favour of one of the two alleles of a gene. For instance, if individuals with allele 'A' are more successful in reproduction than the individuals with 'a', the frequency of the former will be higher. The selection can be artificial or natural. The factors influencing selection may include temperature, humidity, food, sexual attractions, etc.

Harlap and Martini in 1938 conducted an experiment to show the influence of selection. They used ten varieties of barley and mixed equal quantities of seed from each of them. The mixture was grown at five different locations in U.S.A. The experiment was conducted for 4–12 years by continuous harvesting and re-sowing. If there were no selection against or in favour of any variety, one would expect that even after 4–12 years, the proportion of all ten varieties should be equal at each location. Contrary to this when at each location, 500 plants taken at random were classified into the ten varieties, it was found that different varieties predominated at different locations. This is a clear evidence of selection having operated in different locations in favour of different varieties. In view of this, the selection can be regarded as a process of scrutiny which rewards vigorous varieties with reproductive success. There are several aspects of selection, like fitness, gametic selection and zygotic selection.

When one genotype can produce more offspring than the other in the same environment, it means relative reproductive success, called fitness or adaptive value or selective value. For instance, if a genotype carrying 'A' produces 100 individuals reaching maturity and another genotype carrying 'a' produces 90 individuals in the same environment, this would mean that the reproductive success of 'a' is reduced by 10% or by a fraction of 0.1. This adaptive value, which is also designated as 'W', ranges from 1.0 for most reproductive genotypes to zero for lethals. This may also be expressed in the form of selection coefficient or 's' which is defined as a force acting on a genotype to reduce its adaptive value. If the adaptive value 'W' is 1, selection coefficient 'S' would be zero or vice versa. Algebraically, this can be expressed as

$$W = 1 - S \text{ or } S = 1 - W.$$

The rate at which an allele changes in frequency due to selection depends on the intensity of selection and the dominance relations among the genotypes. Under directional selection, dominant allele will increase much more rapidly than recessive alleles, because homozygotes and heterozygotes are favoured. With incomplete dominance, the heterozygotes have selective advantage, so incompletely dominant alleles increase in frequency at a lower rate than that of dominant alleles. Recessive alleles increase at the lowest rate, because only the homozygotes are favoured by selection.

MUTATION AND NATURAL SELECTION

Recurrent mutation and natural selection act as opposing forces on detrimental alleles; mutation increases their frequency and natural selection decreases their frequency. Eventually, these two forces reach an equilibrium, in which the number of alleles added by mutation is balanced by the number of alleles removed by selection.

Achondroplasia can be taken as an example. This is a common type of human dwarfism that results from a dominant gene. Achondroplasic people are fertile and they produce only about 74% as many children as are produced by normal people. Therefore the fitness of people with achondroplasia in average is 0.74, and the selection coefficient,'S' is 1–W or 0.26. If it is assumed that the mutation rate for achondroplasia is about 3×10^{-5} (a typical mutation rate in man), then the equilibrium frequency that can be predicted for the achondroplasia allele is $0.00003/0.26 = 0.0001153$. This frequency is closer to the actual frequency of the disease.

MIGRATION

Through migration, new alleles can be introduced into a population from nearby populations. The overall effect of migration is twofold: 1) It prevents genetic divergence between populations and 2) It increases genetic variation within populations.

Let us consider a large continental population donating individuals (genes) to a small island population. The rate of migration is m, which is equal to the fraction of genes on the island, which are replaced by genes from the continent in each generation. If qi is the frequency of a particular allele 'a', on the island, and qc is the corresponding frequency on the continent, then after migration, the frequency on the island will be

$$qi = (1 - m) \, qi + mqc$$

The change in the frequency of 'a' in one generation will therefore be

$$\Delta qi = qi' - qi = m(qc - qi)$$

where, Δqi will become zero either when migration stops ($m = 0$) or when the frequency of 'a' on the island equals the frequency of 'a' on the continent ($qi = qc$).

The effect of migration is to make populations genetically similar. Even a few migrants per generation are sufficient to eliminate the differences among geographically separated populations. Thus, migration can be a powerful homogenizing force.

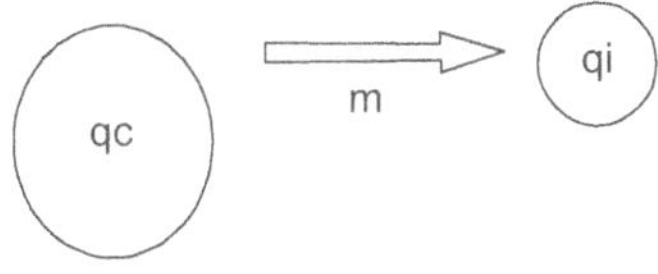

RANDOM GENETIC DRIFT

This was explained by Sewall Wright and called Sewall Wright effect. The Hardy–Weinberg law assumes random mating in a large population. The sample of a population, although representative of the population may not have exactly the same frequency as that found in the population. Such deviations in gene frequencies will be due to sampling errors. This again depends on the size of the sample. The sampling error can arise through founder effect, which is due to the establishment of a population by a small number of individuals. For instance, if only a few parents are chosen to begin a new population, the gene frequency of this new population may deviate widely from the gene frequencies of the original population, because in this case sampling error will be high. On the other hand, if a large number of parents are taken to begin a new population, deviation in the gene frequencies may not be so large.

The deviation in the gene frequency can be measured with the help of standard deviation $\sigma = \sqrt{pq \, / \, N}$, where p and q are the original frequencies of the alleles and N is the number of genes sampled. The number of genes sampled will be equal to double the number of individuals due to diploid condition. Therefore the above formula will take the form $\sigma = \sqrt{pq \, / \, 2N}$, where N is the number of actual parents.

This is illustrated in the following example where $p = q = 0.5$ and the individuals are 1000. Then σ will be $\sqrt{(0.5 \times 0.5) \, / \, 2000} = \sqrt{0.25 \, / \, 2000} = 0.011$. Therefore the gene frequency of the new population will fluctuate around 0.5 ± 0.011. On the other hand if a new population is started by 2 individuals then $\sigma = \sqrt{0.5 \times 0.5 \, / \, 4} = \sqrt{0.0625}$, which means the gene frequency will fluctuate around 0.5 ± 0.25, i.e., –0.25 to 0.75.

In actual practice, the gene frequencies due to random drift may approach the limit that is zero and one. This would be possible only when the new population arises due to a very small sample leading to the fixation of one allele and elimination of the other allele.

Summary

✧　A Mendelian population is a group of interbreeding individuals, whose set of genes constitutes the gene pool.

✧　A population's genetic composition is described by its gene and genotypic frequencies.

✧　Evolution occurs through changes in the gene frequencies. Hardy–Weinberg law describes the effect of reproduction and Mendel's laws on the allelic and genotypic frequencies of a population.

✧　It assumes that a population is large, randomly mating and free from the effects of mutation, migration, selection and genetic drift.

- ✿ When these conditions are met, the allelic frequencies do not change and the genotypic frequencies stabilize after one generation.

- ✿ Applying binomial theorem the gene and genotype frequencies for dominant and recessive alleles can be calculated.

- ✿ If there are three alleles, A_1, A_2 and A_3 at the locus, the trinomial expansion, will represent the genotypic frequencies at the equilibrium stage.

- ✿ For an X-linked locus with two alleles $X^A X^a$, there are five possible genotypes. When an X-linked allele is uncommon, the trait will therefore be much more frequent in males than in females.

- ✿ A number of factors alter the gene equilibrium. Mutation introduces new genes leading to genetic differences in the population.

- ✿ The gene frequencies may change due to selection in favour of one of the two alleles of a gene.

- ✿ Recurrent mutation and natural selection act as opposing forces on detrimental alleles; mutation increases their frequency and natural selection decreases their frequency.

- ✿ Through migration, new alleles can be introduced into a population from nearby population. The overall effect of migration is twofold— it prevents genetic divergence between populations and it increases genetic variation within populations.

- ✿ The gene frequencies due to random drift may approach the limit that is zero and one.

REVIEW QUESTIONS

1. What assumptions must be met for a population to be in Hardy–Weinberg equilibrium?

2. What is the significance of Hardy–Weinberg law?

3. Give the expected genotypic frequencies for

    i. an autosomal locus with three alleles and

    ii. an X-linked locus with two alleles.

4. How does selection alter the allele frequency in a population?

5. What are the effects of mutation and selection on gene frequencies?

6. Explain Sewall Wright effect.

7. The following genotypes were observed in a population

Genotype	Number
BB	50
Bb	45
bb	40

  i. Calculate the allelic frequencies.

 ii. Calculate the genotypic frequencies expected if this population is in Hardy–Weinberg equilibrium.

18

HUMAN GENETICS

Objectives

- ✿ To study the application of genetic principles to humans
- ✿ To learn unique methods of studying human inheritance
- ✿ To understand the basics of genetic counselling

Key Terms

mitogen	colchicine	karyotype
centromere	metacentric	submetacentric
acrocentric	Giemsa stain	G bands
C banding	reverse banding	FISH
pedigree charts	monozygotic twins	dizygotic twins
concordance	discordance	propositus
prenatal diagnosis	invasive techniques	ultrasonography
microcephaly	cleft lip	anencephaly
hydrocephaly	spina bifida	amniocentesis
amniotic fluid	therapeutic abortion	chorionic villi
fetoscopy	endoscope	preimplantation
genetic diagnosis	surgical therapy	cleft palate
polydactyly	hypertension	beta blockers
hydroxyurea	gene therapy	transplant
ADA deficiency	Quinacrine mustard	eugenics
PKU	somatic gene therapy	
alpha-fetoprotein (AFP)	phytohaemagglutinin	

prospective counselling   chorionic villus sampling (CVS)

chromosome banding   retrospective counselling

genetic counselling   hypercholesterolaemia

non-invasive techniques

Severe combined immunodeficiency (SCID) disease

INTRODUCTION

The concepts in Genetics are applicable to all organisms including humans. Human Genetics is a branch of Genetics that deals with the inheritance of traits in man. Man is unsuitable for experimental studies to draw conclusions regarding inheritance pattern and gene expression. In addition social implications must be taken into consideration. In other organisms, geneticists carry out specific crosses to study the inheritance but in humans, controlled matings are not possible. Humans have long generation time and a geneticist should wait approximately for 40 years to observe the F_2 generation progeny. In addition the human family size is small. The genetic studies of humans are highly complex but with all these constraints, the study of human genetics is tremendously important. So unique techniques that suit human biology and culture have been developed to study human inheritance. Several chapters in this book deal with various aspects of human genetics and can be referred.

HUMAN CHROMOSOME COMPLEMENT

The basics of human inheritance are well understood by studying the human chromosomes. The number of chromosomes in a human body cell is 46 ($2n$) and in sperm or ovum is 23 (n). Figure 18.1 shows a clear picture of a diploid set in a somatic cell at metaphase in mitosis.

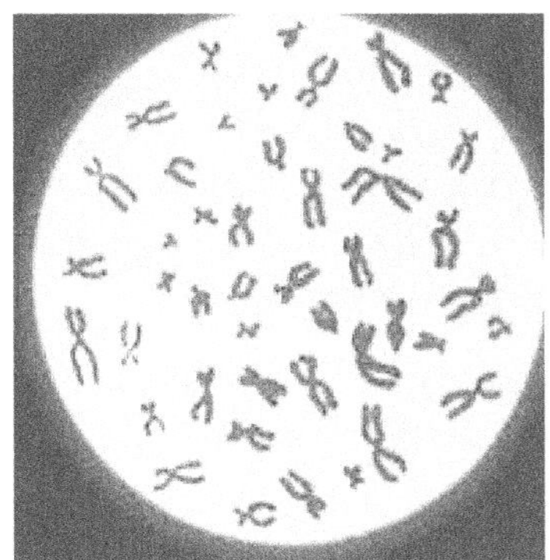

Figure 18.1   Metaphase spread of human chromosomes
from a cell
(*Source*: www.nature.com)

HUMAN CHROMOSOME PREPARATION

Individual chromosomes can usually be identified during cell division. They can be distinguished by size, shape and the patterns produced by various stains. Tijo and Leaven in 1956 devised a unique technique to view human chromosomes from cells. For study, human chromosomes are isolated from white blood cells and stimulated to divide in a culture medium with a mitogen (e.g. phytohaemagglutinin) which increases the rate of cell division. Special treatment with a mitotic inhibitor, e.g. colchicine, stops cell division at metaphase stage, when chromosomes are clear and highly condensed. Treating the cells with hypotonic solution separates the chromosomes. The cells are processed and dropped onto a microslide, and air-dried. The slides are then stained and photographed.

Human Karyotype

The chromosomes of a body cell occur in pairs. For any one chromosome of a particular size and shape, there is usually another match to it and this pair is said to be a homologous pair. When the chromosome set is inspected, it is seen that the length of the chromosome arms vary from one group of chromosomes to another. This is due to variation in the location of the centromere—the constricted region which holds the two arms together.

In some chromosomes the centromere is in the centre, so the two arms of the chromosomes are of equal length. Such a chromosome is called metacentric and the centromere is said to be median in position. If the centromere is slightly off the centre, it is said to be submedian. The chromosome is then classified as submetacentric and will have one arm somewhat longer than the other. If the centromere is located at one end, then one arm is distinctly longer in relation to the other, which may appear tiny. The chromosome is then termed as acrocentric (Table 18.1).

Table 18.1 Classification of human chromosomes

Group	Chromosome number	Chromosome features
A	1, 2, 3	Large, metacentric
B	4 & 5	Large submetacentric
C	6–12 & X	Medium size, submetacentric
D	13, 14, 15	Medium size, acrocentric
E	16, 17, 18	Shorter than D: metacentric or submetacentric
F	19 & 20	Short, metacentric
G	21, 22 & Y	Very short, acrocentric

The chromosomes could be paired and systematically arranged according to the descending order of size and position of centromere and this is termed karyotype. In human karyotype, 7 groups from A to G can be recognized. In the female every chromosome has a matching pair. But in a male one pair does not match. These are the sex chromosomes, a large X and a small Y, distinctly different in appearance. All the chromosomes in a cell other than the sex chromosomes are called autosomes. The autosomes are numbered 1–22 on the basis of decreasing size. A human female thus has 22 pairs of autosomes plus two XX chromosomes. In contrast, a human male cell has 22 pairs of autosomes plus one X chromosome and one Y (Figures 18.2 and 18.3). On the basis of size, the X chromosome is placed in group C, and Y in group G. Chromosomes are numbered based on their descending order of length. The largest chromosome has the lowest number, i.e., 1 and the smallest chromosome the highest number, i.e., 22.

Deviations from the normal karyotype may be associated with abnormalities.

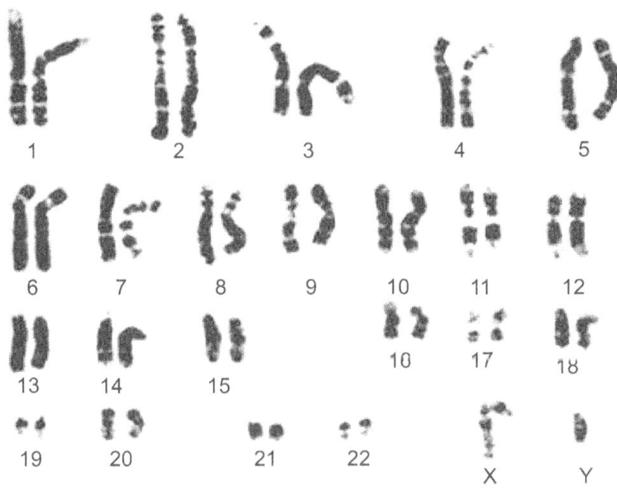

Figure 18.2   Normal male karyotype
(*Source*: www.colarado.edu)

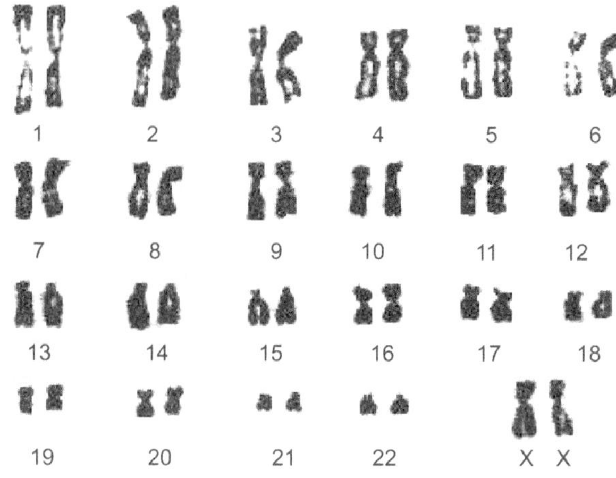

Figure 18.3   Normal female karyotype
(*Source*: www.contexo.info)

Chromosome Banding

The traditional techniques used dyes that stain uniformly and produced uniform colouration in chromosomes. In 1970, Caspersson *et al.* introduced newer staining procedures and different dyes that produce dark and light cross bands of varying widths. Each chromosome in a haploid set has a unique banding pattern and now can be distinguished from other chromosomes of similar size and shape.

Generally there are four kinds of banding such as Giemsa or G banding, centromeric or C banding, Quinacrine or Q banding and reverse or R banding.

G banding   This is a widely used procedure. It involves Giemsa stain after trypsin treatment of metaphase chromosome to differentiate chromosome bands. When viewed under the microscope the chromosomes show dark bands (G bands) and light bands (R bands). By this method approximately 400 dark and light bands could be resolved in a haploid set of chromosomes.

C banding   It requires heating in an alkali solution and staining with Giemsa. C bands are areas of constitutive heterochromatin located adjacent to the centromeres of all the chromosomes.

Q banding   This procedure uses Quinacrine mustard to stain chromosomes. They are observed under fluorescent microscopy. Bright Q bands are equivalent to G bands.

R banding   This is reverse banding. It uses Giemsa under elevated temperature to produce the reverse pattern to that seen in G banding.

Chromosome Painting

In the 1980s, a technique called FISH (fluorescent *in situ* hybridization) was developed to prepare brilliantly coloured whole chromosomes. This technique requires complex mixtures of specially constructed DNA sequences called probes that are complementary to different chromosome regions, together with a set of fluorescent dyes. Also needed is a special microscope with special optical filter sets, as well as computer software to analyse the images and convert them into a dazzling multicoloured display. Chromosome painting is useful for quickly detecting cells with extra or less chromosomes or with other structural abnormalities, especially in cancer cells.

Chromosome Features

Each chromosome has usually two arms held together at the centromere. The short arm is 'p' arm and the long arm is 'q' arm. Regions and bands are numbered from the centromere out. To identify a band in the chromosome, a sequence of 4 items are used namely chromosome number, arm, region and band number. For example, 9q34 refers to chromosome 9, the long arm, region 3, and band 4.

MENDELIAN TRAITS

Refer chapter 4

BLOOD GROUP INHERITANCE

Refer chapter 6

METHODS TO STUDY HUMAN INHERITANCE

Pedigree Charts

Refer chapter 4

Twins Study

The expression of almost all the genes in man is altered by the environment. The effect of environment on the expression of certain genetic traits can be studied by analysing twins. Human females usually give single birth but sometimes two babies are produced at the same time and are called twins. Twins are of 2 types.

    i.   Identical or monozygotic twins (MZ twins)

   ii.   Fraternal or dizygotic twins (DZ twins)

Monozygotic twins They arise from a single zygote formed by fertilization of a single egg with single sperm. This single zygote at the first division gives rise to two identical cells developing independently into two babies. They are generally enclosed in one membrane and are attached to one placenta only. These MZ twins will have identical characters like similar sex and completely resemble each other and are termed concordant.

Dizygotic twins They arise from two different eggs, each fertilized by a different sperm. They are often enclosed in separate foetal membranes and have separate placenta.

They are just like other siblings and are non-concordant type as they may be dissimilar and may be of different sex.

Twins analysis   Francis Galton has introduced the twin study for the analysis of hereditary and environmental factors in human traits. The mean differences in a given trait between the members of a pair of MZ or DZ are computed. If the trait variability is determined by heredity then the intrapair differences will be significantly more among DZ than among MZ. If the trait is controlled purely by the environment, the difference among fraternal and identical twins should be equally expressed (Table 18.2).

The pattern of quantitative inheritance can be represented in a histogram. The intrapair variance of a given trait is compiled as the sum of squares of the difference between the two twins of each pair divided by the number of twin pairs of each kind studied. If the observed variation of a given trait is purely environmental, the intrapair variance for the fraternal twins should be as great as that for the MZ, but less among MZ if the character is controlled genetically.

In MZ, the degree of correlation between the intrapair would be perfect (i.e., correlation coefficient would be 1) but the DZ would show a lower correlation. For a trait controlled by the environment, the correlations for DZ and MZ should be almost the same.

In a pair of twins, if a character is either present or absent in both the members, it is called concordance (++). Identical twins arc always concordant. In a pair of twins, if a character is present in one but is absent in the other, it is called discordance (+–).

A number of human traits have been studied with respect to discordance or concordance. In DZ and MZ the percentage of concordance and discordance with respect to certain traits has been reported in the following table.

Table 18.2   Twins analysis for various diseases

Traits	MZ		DZ	
	++	+–	++	+–
Measles	95	5	87	13
Tuberculosis	74	26	28	72
Diabetes mellitus	84	16	37	63
Tumours	61	39	44	56

CHROMOSOME MAPPING

Refer chapter 8.

DETERMINATION OF SEX

Refer chapter 9.

SEX-LINKED INHERITANCE

Refer chapter 9.

CHROMOSOMAL ANOMALIES– SYNDROMES

Refer chapter 10.

IN BORN ERRORS OF METABOLISM AND OTHER GENETIC DISORDERS

Refer chapter 11.

GENETIC COUNSELLING

Genetic counselling deals with giving advice to families having children with genetic disorders. The aim of genetic counselling is to convey medical and genetic facts to an affected family in a way that can be understood.

Every year thousands of families are affected by the birth of an abnormal child. About 0.7% of newborns have a chromosomal abnormality, another 1% suffers from single gene defects and about 2% have malformations, putting altogether, about 4% newborns have a defect.

With proper genetic counselling, the appearance of serious, sometimes fatal, diseases can be avoided. Such counselling helps to reduce the number of genetically affected families in the population at large and therefore could be enforcing eugenics.

Genetic counselling is effectively done by a team comprising

i. The medical specialist who is specially trained in medicine and is a medical practitioner. He/she plays a consultant role.

ii. The genetic counsellor is a trained person in counselling with a sound knowledge in human genetics. The role of the counsellor is to be involved in the counselling process and in the support and follow-up of the family.

iii. Additional members may be included like social workers, religious support persons and family physicians.

In general, there are two ways by which genetic counselling is effected. They are prospective counselling and retrospective counselling. Prospective counseling is delivered to a person in the reproductive age group before the birth of an affected child. On the other hand, retrospective counselling is rendered after the birth of an affected child.

Steps in Genetic Counselling

Medical diagnosis   Precise diagnosis is required as many hereditary diseases have multiple causes. Some diseases are phenotypically indistinguishable from one another. The propositus (the affected person who gives a clue about the disease) is first examined and in addition other family members are examined. Laboratory work often includes karyotyping; biochemical analyses of blood, urine or cultured cells; or molecular analysis with restriction enzymes and DNA probes.

Pedigree analysis　A complete three- to four-generation family pedigree must be obtained and analysed. The reliability of the collected information must be thoroughly assessed. Some important clues, adoptions, illegitimate births, miscarriages, stillbirths or mildly affected relatives may be entirely missing. Thus decision on the inheritance pattern of the disease must be made, analysing all the facts. Pedigree analysis has been already discussed in chapter 4.

Estimating recurrence risks　The answer to the question, how likely is it that a disease will recur in a subsequent birth? If it is an autosomal recessive allele then the probability is 25% when the parents are known to be heterozygous. For a person heterozygous for an autosomal dominant allele, the chances of any of his children being affected are 50%. In an X-linked trait, the sister of an affected male (because of their mother being a carrier) has 50% chances of being a carrier and therefore the chances of any of her sons being affected are 1 in 4. If a woman has only one son affected, it can be due to a mutation.

Options

1. A couple may refrain from bearing a child, when the genetic disease is definitely going to occur.

2. Prenatal diagnosis can detect the defect before birth and abortion can be advised and the pregnancy is terminated.

3. Alternatively a couple may choose adoption or an appropriate reproductive technology like IVF (*in vitro* fertilization) using donor sperm or donor egg.

Prenatal Diagnosis

The term "prenatal diagnosis" refers to the ability to detect all genetic disorders before the birth of the child. Prenatal diagnosis employs a variety of techniques to determine the health and condition of an unborn foetus. The ability to detect all major chromosomal anomalies and gene-controlled biochemical defects has been a tremendous aid to genetic counselling. Such information can guide the family in reproductive decision-making which may involve terminating the pregnancy, or facilitate planning of appropriate medical, surgical or psychological support.

There are a variety of techniques available for prenatal diagnosis. In general they are classified as

  i.   Non-invasive techniques that do not require an incision or surgery and

  ii. Invasive techniques that often require incision or surgery.

Non-invasive techniques

Ultrasonography   In ultrasound scanning, sound waves of very high frequency are applied. Echoes are reflected from organ boundaries and the degree of reflection depends on the thickness of the tissues. These are electronically transformed into an image in the computer monitor and a sonogram (scan) is developed. This technique is done for all invasive techniques also.

Several anomalies like microcephaly, anencephaly, hydrocephaly, cleft lip, abnormalities of the brain, abdominal organs and heart are usually detectable. There are some limitations to this technique. Some abnormalities like Down syndrome cannot be detected by ultrasonography.

Screening of maternal blood

AFP test   Alpha-foetoprotein (AFP) is a protein produced by the foetus and can be traced in the maternal blood. If abnormally high levels of AFP is detected then it indicates the occurrence of neural tube defects (NTD) in the foetus. The common NTDs are: 1) Anencephaly—failure of the development of the anterior part of the brain, 2) Hydrocephaly—a heavy accumulation of cerebrospinal fluid in the head region, 3) Spina bifida—abnormal development of the lower part of the spinal cord.

Measuring the level of AFP in the mother's blood would detect about 85% of these defects.

Invasive techniques

Amniocentesis   After knowing the foetus position by ultrasound scan and observing in a monitor, the amniotic fluid that surrounds the foetus is drawn. The amniotic fluid consists of foetal cells derived from the skin, respiratory and the urinary systems of the foetus. To obtain the cells, a long, thin needle is inserted through the pregnant woman's abdominal wall, uterus and foetal membranes by a medical practitioner and about 20 ml of the amniotic fluid is withdrawn. This is done often at 14th to 16th week of pregnancy (Figure 18.4). Cells from the sample are then grown in tissue culture and subsequently analysed. Also, the fluid may be subjected to various biochemical tests.

In general the following tests are done.

1.   Biochemical tests to detect enzyme or protein deficiency in inborn errors of metabolism like PKU.

2.   Cells are processed for karyotyping, and chromosome analysis is done for chromosomal disorders.

3. DNA analysis is performed to find out gene mutation and carrier detection.

4. The fluid may be subjected to AFP test.

These procedures may take about 3 weeks and just sufficient time is available for the couples to decide about therapeutic abortion. This procedure is safe in a trained hand. Most women do not report the amniocentesis procedure to be painful. It usually takes a minute or less to perform.

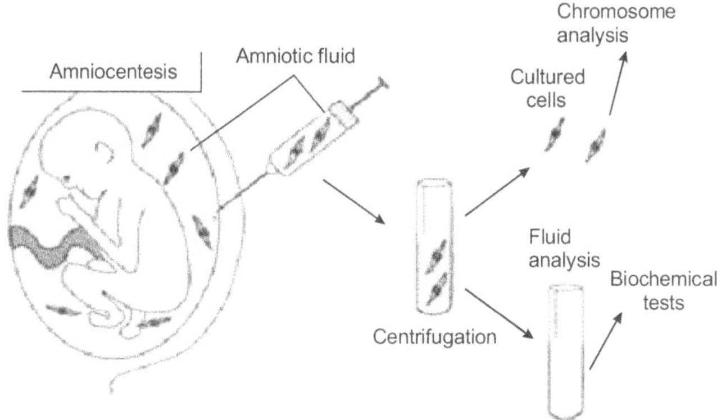

Figure 18.4   Amniocentesis
(*Source*: www.utm.utoronto.ca)

Chorionic villus sampling (CVS)   This procedure was first adopted in China, in 1975, as a means of determining sex of a foetus. It is a specialized alternative test to amniocentesis. Here a biopsy is obtained from the placental tissue or like amniocentesis, a needle is introduced and a bit of foetal tissue from the chorionic villi is sucked off from the placenta. The chorion is the foetal membrane with fingerlike projections namely the chorionic villi on the outer side. Both these are guided by ultrasound sonography. This is done during the 9th to 12th week of pregnancy (Figure 18.5). The sample tissue is subjected to the tests as it is done in amniocentesis.

It is more advantageous than amniocentesis in the following aspects.

i. It is performed in the early periods of pregnancy.

ii. It contains only foetal tissue.

iii. The cells are actively dividing; so chromosome study can be done without cell culture.

iv. The results of the analysis are available much earlier for decision.

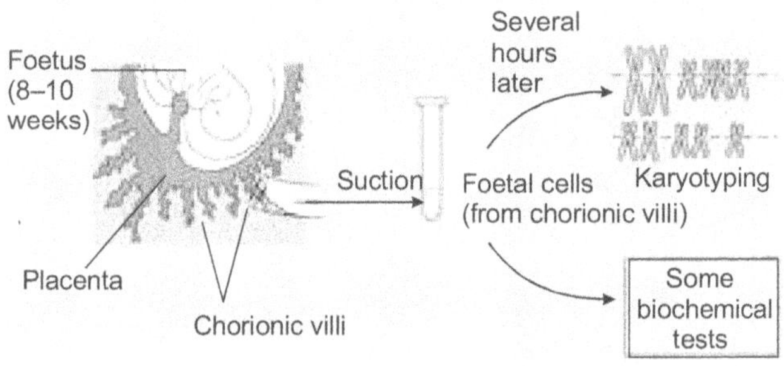

Figure 18.5   Chorionic villus sampling
(*Source*: www.anselm.edu)

Foetoscopy   This is direct visualization of the external features of the foetus with the aid of a fine instrument the endoscope, introduced into the amniotic sac. This is also used to obtain foetal blood samples, skin biopsies, etc. The optimum age is 18–20 weeks of pregnancy and it carries a risk of 3% foetal loss.

Preimplantation genetic diagnosis   In the context of "test-tube baby" (*in vitro* fertilization), one or two cells are collected when the embryo is in the 6–10-celled stage for direct examination.

TREATMENT OF GENETIC DISEASES

In principle, genetically influenced conditions can be treated. But even with the advancements in molecular genetics, only 12% of the genetic diseases could be considered for treatment and the success rate is very low. Some of the better known genetic diseases, like Down syndrome, muscular dystrophy, etc. remain untreatable. It would be best if genetic diseases could be prevented from occurring. Prevention becomes a matter of genetic testing and counselling.

Surgical Therapy

The disfiguring due to a defect may be corrected by surgery. The disfigurement, speech difficulties and swallowing problems of cleft lip and cleft palate can be restored by skillful surgery. The extra digits of polydactyly can be easily removed.

Drug Therapy

Some genetic diseases can be managed by drugs. In hypertension, beta blockers can be used to reduce the pressure on the aorta. Heterozygotes for hyper-cholesterolaemia have accumulation of cholesterol to a high level. This leads to atherosclerosis, that is, hardening of blood vessels and may result in heart attacks beginning from 30 years of age. The symptoms are due to excessive LDL (low-density lipoprotein). A number of drugs are available in the market to reduce the serum cholesterol level. Hydroxyurea is used in the treatment of sickle-cell disease. This drug increases the production of foetal haemoglobin, thus reducing the sickling of red blood cells.

Diet Therapy

The diet therapy for PKU is an example for this type of treatment. A diet low in phenylalanine prevents the accumulation of the amino acid phenylalanine. Avoiding milk which is rich in phenylalanine also reduces the accumulation of phenylalanine.

Replacement of a Missing Gene Product

In diabetes the missing gene product is insulin. Recombinant DNA techniques have been used to make insulin and injection of this insulin to a diabetic will supply the missed insulin.

Similarly in haemophilia A, the missing blood clotting factor viii (antihaemophilic globulin) could be supplied. Though an expensive therapy, biotechnologically produced clotting factor viii is available for treatment.

Organ and Tissue Transplantation

An interesting medical procedure for transferring the correct genetic information into patients who have a hereditary disease is to transplant an organ or tissue from a normal individual. Such a graft may provide a missing enzyme or protein within the patient's body continously.

For example, a "bubble boy" with severe combined immunodeficiency (SCID) disease could receive a bone marrow transplant from his siblings. In SCID, the functioning of B cells and T cells is damaged and so the child is prone to all infections. The bone marrow when transplanted will supply functional B cells and T cells. Bone marrow transplant is also attempted to treat thalassaemia.

Gene Therapy

This is a cure for genetic diseases. A normal gene is supplied which will produce normal protein to do the normal function. Somatic gene therapy is done using body cells to carry the correct gene.

Gene therapy has been successfully performed for ADA deficiency. The first human trial of gene therapy began in

1990. A four-year old girl was given gene therapy (Figure 18.6). Adenosine deaminase (ADA) is an enzyme necessary for the functioning of T cells. The T cells are involved in immunological response to fight against an infection. The deficiency of this enzyme causes immunodeficiency and repeated infections in the child.

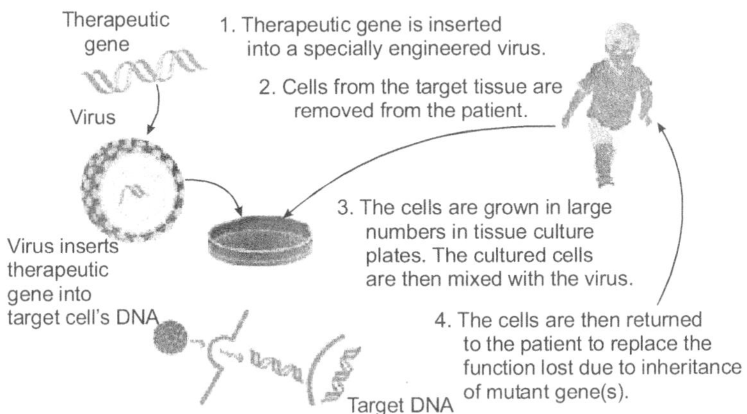

Figure 18.6 Gene therapy
(*Source*: www.history.nih.gov)

The following steps are involved in gene therapy.

1. T cells are isolated from the blood of a patient. Then these cells are cultured in the laboratory.

2. Simultaneously a vector is prepared by inserting ADA gene into the virus. The widely used vector is a retrovirus that is genetically modified. The gene for ADA is artificially synthesized and introduced into the retrovirus genome which becomes the carrier of this gene. The T cells carrying the ADA gene from the retrovirus are the gene-corrected cells.

3. These cells are then returned to the patient by blood transfusion.

Summary

- ○ Human Genetics is a branch of genetics that deals with the inheritance of traits in humans.
- ○ The genetic studies of humans are highly complex and so unique techniques have been developed to study human inheritance.
- ○ In human karyotype, 7 groups from A to G can be recognized.
- ○ The autosomes are numbered 1–22 on the basis of decreasing size.
- ○ Each chromosome has specific banding pattern and several banding techniques are applied to distinguish the chromosomes of similar size and shape.
- ○ Human Mendelian traits, blood group inheritance, pedigree charts, chromosome mapping, sex-determination mechanism, sex-linked inheritance, chromosome anomalies and common genetic diseases have been discussed in previous chapters.
- ○ The effect of the environment on the expression of certain genetic traits can be studied by analysing twins.
- ○ Genetic counselling deals with giving advice to families having children with genetic disorders.
- ○ With proper genetic counselling, the appearance of serious, sometimes fatal, diseases can be avoided.
- ○ Prenatal diagnosis employs a variety of techniques to determine the health and condition of an unborn foetus. Such information can guide the family in reproductive decision-making.
- ○ Management of genetic disorders is carried out by employing various kinds of therapies available.

REVIEW QUESTIONS

1. State the reasons for the difficulties encountered in studying human inheritance.

2. What is the purpose of using the following while preparing human chromosomes for study:

   i.   mitogen

   ii. hypotonic solution

   iii. colchicine

3. Mention the principle applied in classifying human chromosomes.

4. Distinguish chromosome painting from chromosome banding.

5. The concordance value for monozygotic and dizygotic twins for various traits is given below. Indicate the influence of genetic or environmental influences or both for the trait. Explain your reason.

Trait	Concordance rate (%)	
	MZ	**DZ**
ABO blood type	100	50
Smoking	70	70
Schizophrenia	50	20

6. A woman who is pregnant goes to a clinic for genetic counselling. The couple already has a child with Down's syndrome. What procedure could be adopted by the physician while counselling them?

7. Suggest the measures taken while managing genetic disorders.

OBJECTIVE QUESTIONS

Match and choose the correct answer.

1. With reference to phenylalanine metabolism, match the metabolic product from Panel I with the disease in Panel II and choose the correct option.

I	II
1. Phenyl pyruvic acid	a. Alkaptonuria
2. Homogentisic acid	b. Goitrous cretinism
3. Tyrosine	c. Albinism
4. DOPA	d. Phenylketonuria

A. 1d, 2c, 3b, 4a   B. 1c, 2a, 3d, 4b

C. 1b, 2a, 3d, 4c   D. 1d, 2a, 3b, 4c

2. Match the trait in Panel I with the sex-related trait in Panel II and choose the correct option.

I	II
1. Hen feathering	a. Sex-influenced gene
2. Baldness	b. Sex-linked gene
3. Gynandromorph	c. Sex-limited gene
4. Haemophilia	d. Non-disjunction

A. 1d, 2a, 3b, 4c   B. 1c, 2a, 3d, 4b

C. 1c, 2d, 3a, 4b   D. 1d, 2c, 3b, 4a

3. Match the components of *lac* operon in *E.coli* from Panel I with related substance in Panel II and choose the correct option.

	I		**II**
1.	Inducer	a.	RNA polymerase
2.	Promoter	b.	Beta galactosidase
3.	Operator	c.	IPTG
4.	'Z' gene	d.	Repressor
A.	1c, 2d, 3b, 4a	B.	1c, 2a, 3d, 4b
C.	1b, 2a, 3d, 4c	D.	1d, 2c, 3b, 4a

4. Match the cancer-related terms in Panel I and II and choose the correct option.

	I		**II**
1.	Oncogene	a.	Anti-oncogene
2.	Tumour-suppressor gene	b.	Xeroderma pigmentosum
3.	Radiation	c.	Viral genome
4.	Proto-oncogene	d.	*src* gene
A.	1d, 2a, 3b, 4c	B.	1b, 2c, 3d, 4a
C.	1c, 2a, 3d, 4b	D.	1b, 2d, 3a, 4c

5. Match the number of sex chromosomes in Panel I with the number of Barr bodies given in Panel II and choose the correct option.

	I		**II**
1.	XXXY	a.	Three
2.	XO	b.	One
3.	XX	c.	Two
4.	XXXX	d.	None
A.	1d, 2a, 3b, 4c	B.	1b, 2c, 3d, 4a
C.	1c, 2a, 3d, 4b	D.	1c, 2d, 3b, 4a

6. Match the name of the scientist in Panel I with their discoveries in Panel II and choose the correct option.

I	**II**
1. Muller	a. Gene frequency
2. Hardy	b. Mutation detection
3. Curt Stern	c. 21-trisomy in man
4. Lagdon Down	d. Crossing over
A. 1b, 2d, 3a, 4c	B. 1b, 2a, 3d, 4c
C. 1c, 2d, 3b, 4a	D. 1d, 2c, 3b, 4a

7. Match the kind of gene interaction in Panel I with the effect given in Panel II and choose the correct option.

I	**II**
1. Epistatic	a. Gradation in phenotypic expression
2. Complementary	b. Genes affect viability
3. Polygenic	c. One gene masking the effect of another
4. Lethal effect	d. Interaction of two non-allelic genes
A. 1c, 2d, 3a, 4b	B. 1c, 2b, 3d, 4a
C. 1b, 2a, 3d, 4c	D. 1d, 2c, 3b, 4a

8. Match the sex index in Panel I with the sex status given in Panel II and choose the correct option.

I	**II**
1. 0.5	a. Intersex
2. 0.3	b. Super female
3. 1.5	c. Super male
4. 0.6	d. Normal male

A. 1b, 2c, 3d, 4a   B. 1c, 2d, 3b, 4a

C. 1d, 2c, 3b, 4a   D. 1d, 2a, 3b, 4a

9. Match the mutagen in Panel I with their action given in Panel II and choose the correct option.

	I		**II**
1.	Proflavin	a.	Depurination
2.	Nitrous acid	b.	Dimerization
3.	Radiation	c.	Deamination
4.	Alkylating agent	d.	Frameshift

A. 1b, 2d, 3a, 4c   B. 1c, 2a, 3d, 4b

C. 1d, 2c, 3b, 4a   D. 1d, 2a, 3d, 4c

10. Match the genes in the development of *Drosophila* in Panel I with the group to which they belong given in Panel II and choose the correct option.

	I		**II**
1.	Nanos	a.	Homeotic genes
2.	Bicoid	b.	Segmentation genes
3.	Antennapedia complex	c.	Egg-polarity genes
4.	Pair-rule gene	d.	Antero-posterior genes

A. 1b, 2c, 3d, 4a   B. 1d, 2b, 3a, 4c

C. 1c, 2a, 3d, 4b   D. 1c, 2d, 3a, 4b

Read the assertion and reason statements and choose the correct answer from the options given below.

a. Both the Assertion and the Reason are true statements and the Reason is an adequate explanation for the Assertion.

   b. Both the Assertion and the Reason are true statements, but the Reason does not explain the Assertion.

   c. The Assertion is a true statement, but the Reason is a false statement.

   d. Both the Assertion and Reason are false statements.

11. **Assertion:** Colour-blind sons always have colour-blind mother.

    Reason: The inheritance of colour blind gene is holandric gene.

12. **Assertion:** Gynandromorphs are produced by non-disjunction.

    Reason: The cells of the zygote have different chromosome constitution.

13. **Assertion**: Bacterial transformation involves bacteriophage.

    Reason: Phage and bacterium are connected by conjugation tube.

14. **Assertion**: Amniocentesis is advantageous over chorionic villus sampling.

    Reason: Amniocentesis is performed early in pregnancy.

15. **Assertion**: Bacteria transfer genes by transformation.

    Reason: It is the uptake of naked DNA by a bacterium.

16. **Assertion**: Cancer is dangerous.

    Reason: It metastasizes.

17. **Assertion**: Black urine is the symptom of the disease alkaptonuria.

    Reason: The melanin pigment is not produced.

18. **Assertion**: In sickle-cell anaemia, glutamic acid in the haemoglobin, is substituted by valine.

    Reason: The β-globin gene is absent and haemoglobin is synthesized.

19. **Assertion** : Tendency of one crossover interfering with another is called interference.

    Reason: Linkage is the measure of recombination.

20. **Assertion**: Selection introduces new genes in the population.

    Reason: Mutation favours one of the two alleles.

Read the statements and choose the correct answer from the options given.

21. i.   Autosomal dominant inheritance will be shown in 50% of the offspring.

    ii. The males will express in higher frequency than females.

    iii. The trait will be revealed in every generation.

    iv. Example for this pattern of inheritance is albinism.

    v.   The trait is expressed in both the sexes.

    The correct statements pertaining to autosomal inheritance are

    A. i, ii, iii

    B. i, iii, v

    C. ii, iii, v

    D. iii, iv, v

22. The different stages in crossing over is:

    i.   Duplication

    ii. Chiasma formation

iii. Recombination

iv. Synapsis

v. Terminalization

The correct sequence in the crossing over process is

    A. i, iii, ii, v

    B. ii, iii, v, iv

    C. iv, i, ii, v

    D. iii, iv, v, ii

23. The specific features of chorionic villus sampling are:

  i. It is performed when the foetus is 20 weeks old.

  ii. The sample contains foetal cells only.

  iii. It is an invasive procedure.

  iv. It is a therapeutic measure.

  v. It can detect most of the genetic diseases.

The statements **NOT** applicable to chorionic villus sampling are

    A. i, ii, iv

    B. iii, iv, v

    C. i, ii, v

    D. i, iv, v

24. The features of Hardy–Weinberg equilibrium are:

  i. It occurs in a large population.

  ii. The gene frequency fluctuates randomly.

  iii. Mating is selective.

  iv. It is not influenced by mutation and selection.

  v. The gene and genotype frequency is 1.

The **FALSE** statements not applicable to Hardy–Weinberg equilibrium are

    A. ii, iii, iv

    B. i, iii, v

    C. i, ii, iii

    D. iii, vi, v

25. The features of Turner's syndrome are:

  i. 21-trisomy

  ii. Presence of one X chromosome

  iii. Occurrence of epicanthal fold

  iv. Webbed neck

  v. Infertility

    A. i, ii, iii

    B. ii, iv, v

    C. iii, iv, v

    D. i, iv, v

26. Holandric inheritance involves transmission of characters from

  a. male to female   b. female to male

  c. male to male   d. female to female

27. Freemartin is a term that refers to

  a. sterile male calf   b. intersex in man

  c. sterile female intersex calf   d. sex reversed man

28. Haemophilia is a sex-linked disease that occurs

  a. in males only   b. in more males

  c. only in females   d. equally in males and females

29. In the P_1 generation of *Limnaea*, if the mother is sinistral and the father is dextral with regard to shell coiling the offspring of the F_3 generation will be

    a. all dextral   b. all sinistral

    c. 2 sinistral: 2 dextral   d. 3 dextral:1 sinistral

30. In bacterial recombination F prime element transfer is called

    a. conjugation   b. transformation

    c. transduction   d. sexduction

31. The prenatal diagnostic technique that can detect neural tube defects is

    a. radiography   b. AFP test

    c. ultrasonography   d. chorionic villus
                                           sampling

32. Retinoblastoma gene can be categorized as a

    a. tumour-suppressor gene b. proto-oncogene

    c. oncogene   d. viral gene

33. Carcinoma is a term that specifically refers to the cancers of

    a. muscle   b. blood

    c. skin   d. nerve

34. In control of gene expression when repressor binds to operator there is

    a. occurrence of   b. inhibition of
        transcription   transcription

    c. inducer binds to   d. RNA polymerase
        repressor   binds to operator

35. Transduction involves

    a. 2 phages   b. 2 bacteria

    c.   1 bacterium and 1 phage d.   2 bacteria and 1 phage

36. The following is a non-invasive procedure in prenatal diagnosis.

    a.   Amniocentesis   b.   CVS

    c.   Ultrasonography   d.   Endoscopy

37. A diploid organism lacking one chromosome in its genome is

    a.   monosomy   b.   nullisomy

    c.   monoploid   d.   euploid

38. Kappa particle in *Paramecium* is due to the genotype

    a.   KK   b.   KK and Kk

    c.   kk   d.   kk and Kk

39. When a colour-blind man marries a normal woman, among the offspring

    a.   all the males are colour blind   b.   only half of the males are colour-blind

    c.   all the males are normal   d.   half of the females are colour-blind

40. The construction of DNA marker map uses

    a.   microsatellite   b.   phage

    c.   hybrid cell   d.   sequencer

Read the following statements and state whether they are True (T) or False (F).

41. Inversion of chromosome segment is the cause for the formation of acentric and dicentric chromosomes.

42. Polyploids arise by cross breeding between different species.

43. Chromosome painting is performed by fluorescent *in situ* hybridization.

44. Puff induction in a polytene chromosome is a kind of irreversible gene regulation in eukaryotes.

45. Familial Down syndrome is caused by translocation of chromosomes 21 and 14 or 15.

46. Ames test identifies carcinogens by employing mouse.

47. Deletion allows recessive mutations to be expressed causing pseudo-dominance.

48. If the adaptive value 'w' for one allele is 1 then the selection coefficient 's' is also 1.

49. A complementary gene is the second gene that changes the expression of the first gene.

50. The sequence of stages in crossing over is chiasma formation, duplication, synapsis and terminalization.

Keys to the objective questions

1. D	2. B	3. B	4. A	5. D	6. B
7. A	8. C	9. C	10. D	11. d	12. a
13. d	14. d	15. a	16. a	17. b	18. c
19. c	20. d	21. B	22. C	23. D	24. A
25. B	26. c	27. c	28. b	29. d	30. d
31. b	32. c	33. c	34. b	35. d	36. c
37. a	38. b	39. c	40. a	41. T	42. F
43. T	44. F	45. T	46. F	47. T	48. T
49. F	50. F				

GLOSSARY

Achondroplasia A form of dwarfism controlled by a dominant gene.

Acrocentric A chromosome with a terminal centromere making one arm very short.

Adjacent-segregation A type of segregation that takes place in a heterozygote for translocation.

Adenine Purine base in DNA and RNA.

A-DNA Right-handed helical DNA that exists when little water is present.

Albinism A genetic disorder marked by the absence of pigmentation.

Allele frequency Proportion of a particular allele.

Alleles/allelomorphs Alternative forms of a gene found in the homologous chromosome.

Allopolyploidy A polyploid condition where the sets of chromosomes are derived from two or more species.

Alpha-fetoprotein A protein produced by the foetal liver that enters maternal circulation. Its level in the amniotic fluid increases in neural tube defects.

Alpha thalassemia Anaemia caused by a mutation in the alpha globin gene.

Ames test Mutagenicity test in which bacteria are used to evaluate the potential of chemicals to cause cancer.

Amino acid Organic molecule with a carboxyl and amino group, which links to other amino acids to form polypeptide chain.

Amniocentesis A method of obtaining foetal cells from amniotic fluid for prenatal diagnosis.

Amplification The production of multiple copies of DNA sequence.

Anaphase A stage of cell division, during which chromatids of chromosomes move towards opposite poles of the spindle.

Aneuploid Presence of extra chromosome(s) or loss of chromosome(s) in a haploid set of chromosomes.

Antennapedia complex Cluster of genes involved in the development of head and thoracic segments in *Drosophila*.

Antibody A protein (immunoglobulin) produced by the plasma cells in response to foreign substance (antigen).

Anticodon A nucleotide triplet in the tRNA, complementary to the mRNA codon.

Antigen A foreign substance that stimulates the production of antibody.

Antisense RNA Small RNA molecule that pairs with a complementary DNA or RNA and affects its functioning.

Apoptosis Programmed cell death.

Attenuation A type of gene regulation in bacterial operon where transcription is terminated prematurely.

Autopolyploidy A polyploidy condition where the sets of chromosomes are derived from a single species.

Autoradiograph An image produced on a photographic film that was placed on an electrophoresis gel and which shows the positions of radioactive molecules in the gel.

Autosome A non-sex chromosome.

Auxotroph A mutated bacteria that cannot grow in a minimal medium.

Back cross A cross between F_1 individual and one of the parents.

Bacteriophage A virus that infects bacterial cells.

Banding The technique of staining chromosomes to show a pattern of cross bands.

Barr body A dark-stained body representing the condensed, inactivated X chromosome seen in the nuclei of somatic cells.

Base analog A kind of mutagen which can mimic the base pairs.

Base pair A pair of hydrogen-bonded DNA bases located between the two backbones of a DNA double helix.

B cells Lymphocytes that secrete antibody proteins in response to recognizing non-self molecules.

B-DNA Right-handed helical DNA that occurs when water is abundant.

Benign tumour The tumour which does not spread to other parts of the body.

Beta thalassemia Anaemia caused by a mutation in the beta globin gene.

Bioinformatics A discipline formed by combining molecular biology and computer science.

Biotechnology Use of biological processes to produce products of commercial value.

Bithorax complex A group of genes involved in the development of posterior thoracic and abdominal segments in *Drosophila*.

Bivalent A pair of homologous chromosomes as seen during metaphase of the first meiotic division.

Cancer A progressive illness caused by uncontrolled cell division.

Carcinogen A substance that causes cancer.

Carcinogenesis A process of cancer development.

Carrier An individual who is heterozygous for a mutant allele.

Caspases Enzymes that regulate apoptosis.

Catabolite repression Gene control mechanism in bacteria where the metabolism of other sugars is repressed in the presence of glucose.

Cell cycle The sequence of events from one cell division to the other.

Centimorgan An alternate name for map unit.

Centromere The indented region of a chromosome that divides it into two arms.

Chargaff's rule The rules developed by Chargaff regarding the ratios of bases in DNA, i.e., DNA from any cell of all organisms should have a 1 : 1 ratio of pyrimidine and purine bases. Moreover, the amount of guanine is equal to the amount of cytosine and the amount of adenine is equal to that of thymine.

Chiasma A cross-shaped connection between non-sister chromatids during crossing over.

Chorionic villus sampling A method of obtaining foetal cells for prenatal diagnosis, in which a bit of chorionic tissue is removed from the placenta around the 10th week of pregnancy.

Chromatid One of the daughter strand of the replicating chromosome.

Chromosomal aberration An abnormality of chromosome number and/or structure.

Chromosome The hereditary vehicle containing genes and located in the nucleus of an eukaryotic cell.

Cleft palate Failure of the bones forming the roof of the mouth to close.

Clonal evolution A process where mutation accelerates the growth and proliferation of cells in cancer.

Clone A group of genetically identical cells descended from one cell by repeated division.

Codominance Allelic interaction where heterozygotes express the traits of both homozygotes.

Codon A group of three nucleotides in messenger RNA that specifies one particular amino acid.

Complementary bases Nucleotide bases that can pair, purine to pyrimidine, by hydrogen bonding.

Concordant The occurrence of a trait in both twins.

Congenital Any abnormality present at birth.

Conjugation Exchange of genetic material in bacteria by physical contact.

Consanguinity Mating between genetically related individuals.

Cri du chat syndrome The condition resulting from a deletion in the 5th chromosome, and results in a cat cry in the child.

Crossing over The exchange of chromosome parts between the chromatids of homologous chromosomes.

Cystic fibrosis An autosomal recessive disorder caused by the absence of a protein resulting in the formation of thick mucus in the lungs.

Cytoplasmic inheritance Inheritance of characters encoded by genes in the cytoplasm.

Cytosine Pyrimidine base in DNA and RNA.

Deletion Loss of a chromosome segment.

Denaturation Separation of the two strands of DNA in a double helix when DNA is heated.

Deoxyribose The sugar that is present in DNA.

Depurination Loss of purine base in a DNA strand.

Diabetes mellitus Failure of production of insulin leading to a high level of glucose in the blood.

Dicentric chromatid A chromatid that has two centromeres.

Dihybrid An individual heterozygote for two genes.

Dioecious Separate male and female sex.

Diploid Two sets of chromosomes present in somatic cells.

Discordant A pair of twins where one of them has the trait considered and the other twin does not.

Dizygotic twins Non-identical or fraternal twins that arise from two zygotes.

DNA Deoxyribonucleic acid, the genetic material.

DNA fingerprinting A technique for identifying individuals by using DNA probes in electrophoresis.

DNA hybdridization A technique where a complementary base pair is uscd to find out the similarity of genome.

DNA ligase An enzyme that seals two DNA fragments.

DNA polymerase An enzyme that forms a new DNA.

DNA repair A process in which bases that are incorrectly introduced in the DNA are changed to produce the original sequences.

DNA replication Duplication of DNA double helix by using a parental strand.

Dominant gene The allele or phenotype that is expressed even if one allele is present.

Dosage compensation The equalization of expression of X-linked genes in both sexes.

Down syndrome A group of abnormal characters that is seen in an individual due to the presence of three copies of chromosome 21.

Duchenne muscular dystrophy Progressive

muscle weakness caused by a sex-linked recessive gene.

Duplication An extra segment of chromosome or DNA, resulting in excess dose of genes.

Edward syndrome A group of abnormal characters that is seen in an individual due to the presence of three copies of chromosome 18.

Egg-polarity genes Genes that determine the axes of development in *Drosophila*.

Embryo The stage of prenatal development in which organs develop from a three-layered organization.

Empiric risk Computation of probability that a trait will occur in a family based on past experience rather than on knowledge of the trait.

Enhancer The DNA sequences that increase the rate of transcription by interacting with promoters.

Enzyme A protein molecule that in small amounts accelerates the rate of chemical reaction.

Episome Plasmid capable of integrating into a bacterial chromosome.

Epistasis Gene interaction in which gene at one locus suppresses the effect of a gene at another locus.

Eugenics Improvement of human race by encouraging the matings of people with beneficial genes (positive eugenics) and discouraging the matings of people with harmful genes (negative eugenics).

Eukaryote A cell or organism that contains a membrane-bound nucleus with chromosomes.

Euploid The chromosome number that is the exact multiple of a haploid number.

Evolution Changes in gene frequency in a population over a long period of time that may result in new species.

Exon An expressed coding sequence which is a part of split genes.

Expanding trinucleotide repeat Mutation in which the number of trinucleotides increases in succeeding generations.

Favism Anaemia that occurs when a person deficient in G6PD enzyme eats fava beans or takes certain drugs.

Fertilization Fusion of male and female gametes in sexual reproduction.

Fetoscopy The direct visualization of the foetus in prenatal diagnosis by fetoscope.

F factor Episome of *E. coli* that controls conjugation between the bacterial cells.

F_1 generation The first generation of offspring from a mating between two parents.

Fingerprint technique A method of combining electrophoresis and chromatography to separate the components of a protein.

Fitness Reproductive success of a genotype compared with other genotypes in the population.

Fluorescent *in situ* hybridization (FISH) A technique in which fluorescently labelled DNA probes are hybridized with the complementary target sequence.

Founder effect In a small population, sampling error occurs when a population is established by a small number of individuals leading to genetic drift, this is known as founder effect.

Frameshift mutation A mutation where the reading frame of the gene changes due to deletion or addition of bases.

Fraternal twins Non-identical twins that are formed when two different eggs are fertilized by two different sperms.

Gamete A reproductive cell (sperm or ovum).

Gamete intrafallopian transfer (GIFT) Reproductive technology in which sperms are transferred from the laboratory to a woman's fallopian tube, where fertilization occurs.

Gene A DNA sequence that codes for the synthesis of a protein.

Gene amplification A technique that uses enzymes for DNA replication *in vitro* to make many copies of a particular DNA sequence.

Gene flow Movement of alleles between populations.

Gene frequency The proportion of genes in a given population.

Gene interaction Interaction between genes at different loci that affect the same trait.

Gene pool All the genes present in a population.

Generalized transduction Transfer of any gene from one

bacteria to another through a virus.

Gene regulation Process that controls the expression of a gene.

Gene therapy Treating a genetic disease by inserting the normal gene.

Genetic code The triplet bases of DNA or mRNA that specify the different amino acids.

Genetic counselling Providing guidance to a family regarding the occurrence of a genetic disease and ways to deal with them.

Genetic drift Random fixation of genes in a small population.

Genetic map Maps constructed on the basis of recombination rate of genes located in the chromosomes.

Genetic marker Any gene that helps in the identification of a location in the genetic map.

Genetic screening Examining a population for a disease.

Genome The full set of genes in an individual.

Genotype The genetic constitution of an individual.

Glucose 6-phospate dehydrogenase (G6PD) An enzyme involved in aerobic respiration.

Guanine Purine base in DNA or RNA.

Gynandromorph An individual that is a mosaic for sex chromosomes and possesses tissues with different chromosomal constitution.

Haploid The single set of chromosome.

Hardy–Weinberg law In a large randomly mating population, the gene and genotype frequencies remain constant when there is no migration, no mutation, no selection and no genetic drift.

Heat-shock proteins Proteins produced by cells in response to heat and other stresses.

Hemizygous One allelic condition in a male with regard to X-linked genes.

Haemophilia The failure of blood clotting due to the absence of blood-clotting factors.

Heredity Transmission of genetic information from parents to offspring.

Heterogametic sex The sex that produces two different

kinds of gametes with regard to the sex chromosomes they carry.

Heterozygous The occurrence of two different alleles for a trait present in the same locus in the homologous chromosomes.

Holandric inheritance The inheritance of genes present in the Y chromosome.

Homeotic complex Major cluster of genes involved in the development of *Drosophila*.

Homogametic sex The sex that produces one type of similar gametes with regard to sex chromosomes.

Homologous chromosomes The paired similar chromosomes, one of the pair being contributed by the father and the other by the mother during fertilization of gametes.

Hormone A biochemical manufactured in a gland and transported in the blood to a target organ and producing a specific effect.

Homozygous An individual that possesses two identical alleles for a trait.

Hybrid The offspring of a cross between two genetically different organisms.

Hydrocephalus Swelling of the brain by the accumulation of cerebrospinal fluid.

Hypertension Elevated blood pressure caused by the interactions of genes and environmental factors.

Hypostatic gene A gene that is masked by another gene present at a different locus.

Identical twins Twins that arise when the fertilized egg splits and develop into two embryos.

Inborn error A biochemical disorder that is genetically determined, and in which absence of a specific enzyme produces a metabolic block.

Incomplete penetrance The expected phenotype is not expressed by the genotype.

Independent assortment Mendel's law stating that alleles at different loci on a homologous chromosome are transmitted independently from generation to generation.

Intercalating agents Chemical substances that are similar to nucleotides and get inserted between the adjacent bases in DNA.

Interphase A part of the cell cycle where there is no cell division.

Interference Degree to which one crossover interferes with additional crossovers.

Intron Intervening sequences that are parts of a gene alternating with exons.

Inversion A type of chromosomal aberration in which a part of a chromosome is broken and reinserted in the reverse order.

IVF (*In vitro* fertilization) Alternative reproductive technology where the sperm and ovum are allowed to fuse in a glass tube in the laboratory.

Karyotype The orderly arrangement of a chromosome set of a somatic cell.

Kilobase A unit that contains one thousand DNA base pairs.

Klinefelter syndrome An extra X chromosome in a male resulting in abnormalities.

Law of segregation Alleles separate during meiosis.

Lesch–Nyhan syndrome Absence of the enzyme HGPRT (Hypoxanthine/guanine phosphoribosyl transferase) that causes the failure of recycling of the nitrogenous bases of DNA.

Lethal allele An allele that causes death of an individual.

Linkage The association between genes present in the same chromosome so that they are inherited together.

Locus The position or site of a gene on a chromosome.

Lymphocyte A type of white blood cell produced in the bone marrow that provides immunity.

Lyon hypothesis The hypothesis proposed by Mary Lyon and that states that one X chromosome in each female cell becomes inactivated at random and present as Barr body.

Lysogenic cycle Life cycle of a bacteriophage in which the phage genes first integrate into the bacterial chromosome and are then transcribed.

Lytic cycle Life cycle of a bacteriophage in which new phages are produced and the host cell is lysed.

Malformation Morphological defect due to abnormal development.

Malignant A tumour that is capable of spreading to other tissues in the body.

Map unit A unit to measure the distances on a genetic map; 1 map unit is equal to 1% of recombination.

Meiosis A special type of cell division occurring in gametes resulting in the reduction of chromosome number.

Mendelian population Group of interbreeding, sexually reproducing individuals.

Messenger RNA (mRNA) The RNA which has comple-mentary sequences for a DNA strand, is formed during transcription and functions during translation of a poly-peptide.

Metacentric A chromosome with a centrally placed centromere.

Metaphase The stage of cell division where the chromosomes occupy the equatorial plane and have two chromatids each.

Metastasis The spread of malignant tumour cells from one site to other parts of the body.

Mitosis Somatic cell division resulting in the formation of two daughter cells, each with the same chromosome complement as the parent cell.

Monoecious Presence of both male and female reproductive structures in the same animal.

Monohybrid cross A cross between individuals that differ in a single character.

Monosomy The condition where there is loss of a chromosome in a set of chromosomes.

Monozygotic twins Twins derived from a single fertilized ovum.

Mosaicism Single individual having different chromosome constitution in the tissues.

Multiple alleles More than two alleles present in a locus of a chromosome controlling a trait.

Mutagen A substance that induces mutation.

Mutant An individual bearing a mutation.

Mutation A process in which a gene or a chromosome undergoes changes.

Mutation rate Frequency with which a gene changes from a wild type to a mutant.

Natural selection Differential reproduction of genotypes.

Negative control In gene regulation, the binding of a regulator protein to DNA inhibits transcription.

Neural tube defect A birth defect in which a part of the spinal cord protrudes due to the failure of closure of the neural tube.

Non-disjunction Failure of separation of homologous chromosomes during cell division.

Nucleoside A purine or pyrimidine base attached to a ribose or deoxyribose sugar.

Nucleosome A globular unit in a chromatin fibre with a histone core around which the DNA strand is wound.

Nucleotide A nucleoside attached to a phosphate group.

Nullisomy Absence of both chromosomes of a homologous pair.

Oligonucleotide DNA sequence made up of a small number of sequences.

Oncogene A gene that can transform a normal cell into a cancer cell.

Oocyte The female sex cell before it is fertilized.

Operator gene The gene that switches on the structural genes.

Operon A unit of gene function, consisting of an operator gene and structural genes whose action it controls.

Ovaries The paired female reproductive organs which produce eggs.

Ovulation Release of oocytes from the ovary.

Ovum An egg cell.

Pair-rule genes Set of segmentation genes in *Drosophila* that is responsible for the development of regional sections of the embryo and affect alternate segments.

Pangenesis Early concept of heredity that states that particles carry genetic information from different parts of the body to the reproductive organs.

Paracentric inversion Chromosome inversion that does not include centromere in the inverted region.

Patau syndrome Abnormal features that results from the presence of three copies of chromosome 13.

Pedigree A chart or schematic diagram representing the family history of an individual.

Penetrance Percentage of individuals that express the expected phenotype.

Pericentric inversion Chromosome inversion that includes the centromere in the inverted region.

Pharmacogenetics The area of biochemical genetics dealing with genetics of responses to drugs.

Phenocopy The alteration of the phenotype by environmental factors during development which mimics the one that is produced by a gene.

Phenotype The appearance of an individual.

Phenylketonuria (PKU) An inborn error of metabolism in which the absence of an enzyme phenylalanine hydroxylase causes impairment in phenylalanine metabolism.

Philadelphia chromosome An aberrant chromosome with a deletion in the long arm of chromosome 22 resulting in chronic myeloid leukaemia.

Phocomelia Failure of the complete development of limbs.

Physical map Map of physical distance between loci.

Placenta A specialized organ that connects the developing foetus and the mother.

Plasmid Small circular DNA present in bacterial cells that can replicate autonomously.

Pleiotropy Many sided effect of a single gene.

Polydactyly An autosomal dominant trait characterized by extra fingers and toes.

Polygenes Genes that are present in different loci controlling a trait.

Polymerase chain reaction (PCR) A technique that is used to amplify the DNA segment.

Polymorphism The occurrence of two or more forms in the same habitat.

Polypeptide A chain of more than one peptide.

Polyploid Multiples of the basic haploid set of chromosomes.

Polytene chromosome Giant chromosome in the salivary gland of dipteran larvae.

Population genetics Branch of genetics dealing with genetic variation and evolution of a population.

Position effect The expression of genes influenced by their location.

Positive control Gene regulation in which binding of a regulator protein stimulates transcription.

Preformationism Early concept of inheritance proposing that a miniature adult (homunculus) resides in the gametes.

Pre-implantation genetic diagnosis Selecting an embryo produced by *in vitro* fertilization and testing it before implanting into the uterus.

Prenatal diagnosis Techniques employed to diagnose the genetic disease in a foetus before birth.

Primer An oligonucleotide sequence that flanks either side of the DNA to be amplified by PCR.

Proband or propositus An affected individual who is responsible for identifying a disease in the family.

Probe An artificially synthesized labelled DNA or RNA sequence used to detect the complementary sequence.

Progeria An inherited premature aging disease.

Promoter A DNA sequence that initiates transcription process.

Prophase The first phase of cell division in which the chromosomes become visible.

Proto-oncogene A normal gene which when altered can become an oncogene causing cancer.

Pseudodominance Expression of a normally recessive allele due to a deletion in the homologous chromosome.

Punnett square A shorthand method of determining the outcome of a genetic cross.

Purine A type of nitrogenous base in DNA and RNA. Adenine and guanine are purines.

Pyrimidine A type of nitrogenous base in DNA and RNA. Thymine, cytosine and uracil are pyrimidines.

Reading frame The starting DNA base from which a polypeptide sequence is read.

Recessive A trait which is expressed only in homozygous condition.

Reciprocal cross Cross in which the phenotypes of the female and male parents are reversed.

Reciprocal translocation Reciprocal exchange of segments between two non-homologous chromosomes.

Recombinant DNA technology Set of molecular techniques for isolating, manipulating and combining DNA segments.

Recombination Genetic material is broken down and then combined to form new combination of genes. This leads to offspring having newer combination of genes from their parents. It also refers to the artificial and delibrate recombination of disparate pieces of DNA from different organisms to produce recombinant DNA.

Receptor A molecule in or on a cell membrane that has a site to fit another molecule triggering chemical activity in a cell.

Regulator gene A gene which controls the rate of synthesis of a product.

Replication The process that synthesizes new double-stranded DNA.

Restriction endonuclease A bacterial enzyme which cleaves the DNA at particular site.

Restriction fragment length polymorphism (RFLP) Polymorphism seen in DNA sequence in a population due to the presence or absence of restriction site.

Restriction site A DNA sequence that is cleaved by a specific restriction endonuclease.

Retinoblastoma A rare eye cancer in children.

Reverse transcriptase An enzyme that can transcribe RNA into DNA.

Ribonucleic acid (RNA) Single-stranded nucleic acid having ribose sugar.

RNA polymerase An enzyme that binds to a promoter site and synthesizes mRNA from DNA.

Sampling error Deviations from expected ratios due to chance.

Segmentation genes Cluster of genes in *Drosophila* that control the differentiation of the embryo into individual segments.

Segment-polarity genes Segmentation genes in *Drosophila* that control the organization of segments.

Selection coefficient A measure of the intensity of selection against a genotype, which equals 1 minus fitness.

Sex chromosomes Chromosomes that differ in number or morphology in males and females.

Sex-influenced trait A trait whose expression is influenced by the sex.

Sex-limited trait Trait expressed by autosomal genes only in one sex.

Sex-linked trait Characteristics determined by genes in the sex chromosomes.

Somatic cell hybridization Fusion of different cell types.

Southern blot A technique where DNA fragments run on an electrophoresis are transferred to a nylon membrane and detected by hybridization.

Specialized transduction Transduction in which genes near specific sites on the bacterial chromosomes are transferred from one bacterium to another by a virus.

Spermatogenesis The process of formation of sperms.

***SRY* gene** A gene in the Y chromosome that determines maleness.

Structural gene DNA sequence that encodes a protein that functions in metabolism.

Submetacentric A chromosome having centromere shifted from the centre.

Suppressor mutation A mutation that suppresses the effect of other genes.

Synapsis Close pairing of homologous chromosomes.

Syndrome A group of symptoms that occur, characterizing a disease.

Tautomer Two forms of a chemical.

***TDF* gene** The gene located in the Y chromosome in a male that triggers male development.

Telomere The tip of a chromosome.

Telophase The last phase of cell division resulting in the formation of two daughter cells.

Teratogen A chemical or other environmental agent that causes a birth defect.

Test cross Crossing an individual of unknown genotype to a homozygous recessive individual.

Testicular feminization A male embryo developing female organs though they have XY chromosomes.

Thymine Pyrimidine base in DNA.

Totipotent Potential of a cell to develop into any other cell type.

Transcription The formation of mRNA against a DNA template.

Transduction Gene exchange between bacteria through a virus.

Transformation A mechanism in which DNA from the medium is taken up by a cell.

Transition A point mutation altering a purine to purine or a pyrimidine to pyrimidine.

Translation The formation of a protein directed by a specific mRNA.

Translocation A chromosomal aberration involving transfer of a piece of one chromosome to a non-homologous chromosome.

Transversion A point mutation altering a purine to a pyrimidine and a pyrimidine to a purine.

Trihybrid cross A cross between two individuals that differ in three characteristics.

Triplet code A unit of three linear successive bases in DNA or RNA which codes for a specific amino acid.

Triple test A prenatal test that analyses three substances—serum levels of AFP, estriol and beta hCG.

Trisomy 21 An extra copy of chromosome 21, causing Down syndrome.

Tumour-suppressor gene A gene whose products control normal cell growth and a mutation of which leads to cancer.

Turner syndrome The condition in which a female has one X chromosome.

Ultrasonography A procedure where ultrasound waves are used to create an image on a video screen. This is often employed in prenatal diagnosis.

Uracil Pyrimidine base in RNA.

Uterus The muscular, saclike organ in the human female in which the foetus develops.

Virulent phage A bacteriophage that reproduces by lytic cycle and kills the bacteria.

Wild type A allele that is most common for a certain gene in the population which produces a normal phenotype.

Wilms tumour   A childhood cancer of the kidneys.

Wilson's disease   A disease characterized by the inability to metabolize copper.

Xeroderma pigmentosum   A disorder in which DNA repair is impaired and exposure to sun causes mutations and skin cancer.

X-inactivation   The turning off of one X chromosome in a female.

Zygote   The diploid cell resulting from the union of sperm and ovum.

REFERENCES

Anthony, J.F., William G.M., Richard, L.C. and Jeffrey M.H. (2002). *Modern Genetic Analysis*, 2nd edn. W.H. Freeman and Co.

Benjamin, L. (2006). *Essential Genes*. Prentice Hall.

Bhatnagar, Kothari and Mehta. (2002). *Essentials of Human Genetics*. Orient Longman Ltd.

Bruce, K.R. (2007). *Human Genetics and Genomics*, 3rd edn. Blackwell Publishers.

Daniel, H.L. (1996). *Essential Genetics*. Jones and Bartlet Publishers, UK.

David, L.S. (2005). *Gene Regulation* (Bios-Advanced text), 5th edn. Cromwell Press, UK.

Elanie Johansen Mangae and Arthur, P. Mangae. (1999). *Basic Human Genetics*, 2nd edn. Sinaeur Associates.Inc., Sunderland, USA.

Gardner, R.J.M. and Sunderland, G.R. (1996). *Chromosome Abnormalities and Genetic Counseling*. Oxford University Press.

Griffiths, A.F.J., Miller, H.J., Suzuki, T.D., Lewontin, R.C. and Gelbart, M.W. (1993). *An Introduction to Genetic Analysis*. W.H. Freeman and Company, NY.

Gustavo Maroni. (2001). *Molecular and Genetic Analysis of Human Traits*. Blackwell Science Ltd.

Holland, B. and Kyriacou, C. (1993). *Genetics and Society*. Addison-Wesley.

Jennifer Gregory. (2001). *Applications of Genetics*. Cambridge University Press.

John Ringo. (2004). *Fundamental Genetics*. Cambridge University Press.

Knudson Alfred, G. (1971). "Mutation and cancer". *Pro.Natl.Acad.Sci.* Vol. 68.

Morgan, T.H. (2001). *Embryology and Genetics*. Agrobios, India.

Pierce, A. Benjamin. (2003). *Genetics—A Conceptual Approach*. W.H. Freeman and Company, NY.

Scott Hawley and Michelle,W.Y. (2004). *Advanced Genetic Analysis—Finding Meaning in a Genome*. Blackwell Publishers.

Tamarin, R.H. (1993). *Principles of Genetics*. 4th edn. WCB. Oxford.

Ursula Goodenough.(1985). *Genetics*. Holt Reinhart and Winstan, NY.

William, K.S. and Michael, C.R. (2000). *Concepts of Genetics*. Prentice Hall.

INDEX

H

I

Super male 122
Supplementary genes 65
Suppressor 44
Suppressor mutators 150
Surgical therapy 311

T

Tandem duplication 167
Tautomeric shifts 151
Tautomers 151
Tay-Sachs disease   189
Telomerase 210
Telomeres 210
Temperate phage 261
Terminal deletion 165
Test cross 30
Testicular feminization 196
Testis-determining pathway 130
Tetraploids 178
Thalassemia major 192
Thalassemia minor 192
Thalassemias 192
Thalidomide tragedy 44
Thiogalactoside
    transacetylase 220
Three-point test cross 106
Thumb crossing 41
Thymidine kinase 112
Thymine 10
Tobacco mosaic virus 20
Toll 242
Tongue rolling 40
Totipotent 238
Transacetylase 220
trans-arrangement 92
Transducing phage 261
Transformants 261

Transformation 16
Transforming principle 18
Translocation 168
Transversions 155
Trihybrid cross 35
Trihybrid test cross 106
Triploids 178
Trisomic 173
Tumorous head 276
Turner syndrome 173
Two-point test crosses 105
Tyrosinosis 188

U

Ultrasonography 307
Universal donors 82
Universal recipients 82
Uracil 10

V

Variable nucleotide tandem
    repeats 110

W

Wedging protein 261
White eye 79
Widow's peak 40

X

Xeroderma pigmentosum 163
X-inactivation centre 133
X-linked genes 51

Z

Z-DNA 22